# SpringerBriefs in Mathematics

**SpringerBriefs in Mathematics** showcases expositions in all areas of mathematics and applied mathematics. Manuscripts presenting new results or a single new result in a classical field, new field, or an emerging topic, applications, or bridges between new results and already published works, are encouraged. The series is intended for mathematicians and applied mathematicians.

More information about this series at http://www.springer.com/series/10030

# SBMAC SpringerBriefs

The **SBMAC SpringerBriefs** series publishes relevant contributions in the fields of applied and computational mathematics, mathematics, scientific computing, and related areas. Featuring compact volumes of 50 to 125 pages, the series covers a range of content from professional to academic.

The Sociedade Brasileira de Matemática Aplicada e Computacional (Brazilian Society of Computational and Applied Mathematics, SBMAC) is a professional association focused on computational and industrial applied mathematics. The society is active in furthering the development of mathematics and its applications in scientific, technological, and industrial fields. The SBMAC has helped to develop the applications of mathematics in science, technology, and industry, to encourage the development and implementation of effective methods and mathematical techniques for the benefit of science and technology, and to promote the exchange of ideas and information between the diverse areas of application.

http://www.sbmac.org.br/

Barbara Amaral • Marcelo Terra Cunha

# On Graph Approaches to Contextuality and their Role in Quantum Theory

Barbara Amaral
Department of Statistics, Physics
and Mathematics
Federal University of São João del-Rei
Ouro Branco, Minas Gerais, Brazil

International Institute of Physics,
Federal University of Rio Grande do Norte
Natal, Rio Grande do Norte, Brazil

Marcelo Terra Cunha
IMECC – Department of
Applied Mathematics
University of Campinas
Campinas, São Paulo, Brazil

ISSN 2191-8198　　　　　　　ISSN 2191-8201　(electronic)
SpringerBriefs in Mathematics
ISBN 978-3-319-93826-4　　　ISBN 978-3-319-93827-1　(eBook)
https://doi.org/10.1007/978-3-319-93827-1

Library of Congress Control Number: 2018946783

This Springer imprint is published by the registered company Springer Nature Switzerland AG.
The registered company address is: Gewerbestrasse 11, 6330 Cham, Switzerland

# Preface

This book focuses on two of the most striking features of quantum theory—contextuality and nonlocality—fascinating, intrinsically quantum phenomena with both fundamental and practical implications. We are specially interested in graph approaches to contextuality, which were developed from the perception that the knowledge in graph theory could be applied directly to solve several different open problems. This led to a breakthrough in the field, providing new tools and insights, some of them borrowed from other applications of graph theory.

One of the main contributions of the graph approach to contextuality comes from the fact that many sets of probability distributions in contextuality scenarios can be described using well-known convex sets from graph theory, leading to a beautiful geometric characterisation of such sets. This brings into play the tools of convex optimisation, linear programming and semidefinite programming, that can now be used to solve tasks that we did not know how to tackle before the graph approach was developed.

The connection of contextuality and graphs helps us with many practical problems, but its contributions can be even deeper. Many results regarding graph invariants, such as the Lovász theta function, lead to important developments in foundations of quantum physics. This was the main problem discussed in the PhD thesis of one of us, supervised by the other, including important contributions by Adán Cabello and an internship with Andreas Winter (two of the founding fathers of graph approaches to contextuality), the starting point for the book that you are reading now.

In the following pages, you will find our attempt to present the main contributions of graph theory to the study of contextuality and how they help us to understand the subtleties of quantum theory.

We are indebted to many friends for their interest and help with this project. Let us start by Carlile Lavor, who invited us for writing the book, and Adán Cabello, who started our interest for the field. Many colleagues and students were essential, for many different reasons, and we take this opportunity to thank them: Ernesto Galvão, Matthias Kleinmann, Raphael Drumond, Roberto Baldijão, Rui Soares Barbosa, Saulo Moreira and Steve Walborn made valuable comments on previous versions of the text. Shane Mansfield has not only suggested important changes to the text but also sent us the beautiful bundle diagrams, which appear on Sect. 2.6.1. Cristhiano Duarte, for discussions regarding the book, graphs, football, beer, the good and the bad of life in academia. . . Ana Belén Sainz, for endless discussions that helped us understanding much better our own work (and ourselves). Rafael Chaves and all the International Institute of Physics for support and hospitality in Natal.

Agencies, groups and networks always play important roles in science. We are indebted to Capes, CNPq and INCT-IQ for support. EnLight, MathFoundQ and QiQm were essential in bringing the appropriate environment for scientific work. We do thank our colleagues that are part of them! We also thank SBMAC and Springer for this opportunity.

Another part of life was also essential during this process of writting, revising, writting. . . For keeping us reasonably sane, we thank our friends from wherever and specially our families (dogs included!).

Natal, Brazil                                                                                    Barbara Amaral
Barão Geraldo, Brazil                                                              Marcelo Terra Cunha
February 2018

# Contents

# Chapter 1
# Introduction

Quantum theory provides a set of rules to predict probabilities of different outcomes in different experimental settings. While it predicts probabilities which match with extreme accuracy the data from actually performed experiments, it has some peculiar properties which deviate from how we normally think about systems which have a probabilistic description. Two of the "strange" characteristics of quantum theory are *contextuality* and *nonlocality*. The former tells us that we cannot think about a measurement on a quantum system as revealing a property which is independent of the set of measurements we choose to make. The latter describes how measurements made by spatially separated observers in a multipartite quantum system can exhibit extremely strong correlations. Contextuality and nonlocality are probably the most striking features of quantum theory. We believe that a complete understanding of these features may be the most important step towards *understanding* the whole theory.

These "strange" properties of quantum theory are related to the fact that it has an intrinsic statistical character. In contrast with classical theory, quantum theory does not provide the exact values of all measurements, but rather the probabilities of the occurrence of each possible outcome. This happens even when the state of the system is an extremal point of the set of all possible states, a *pure* state, which corresponds to the situation in which we have the best possible knowledge about the system.

Consider a set with a huge number of copies of the same system, all prepared in the same way. Such a set will be called an *ensemble*. To estimate the probability distribution of a given measurement for this preparation, one can perform this measurement in several copies, and count the relative frequencies of each outcome. For most measurements, at least two outcomes have probability larger than zero. Two possible explanations for this indeterminacy on the outcomes of the measurements are *a priori* conceivable:

© The Author(s), under exclusive licence to Springer Nature Switzerland AG 2018
B. Amaral, M. Terra Cunha, *On Graph Approaches to Contextuality and their Role in Quantum Theory*, SpringerBriefs in Mathematics,
https://doi.org/10.1007/978-3-319-93827-1_1

I. Our description of the preparation procedure is incomplete, which allows for different states to be consistent with it. Due to this incompleteness, it is possible that the individual systems of the ensemble are in different states, in such a way that we could separate the copies in a number of sub-ensembles, each of them consisting of a definite state that has definite outcomes for all measurements. The probabilistic character of the experiments is, in this case, explained by our lack of information: we do not know everything about the system we are measuring and hence we cannot predict the results with certainty.

II. All individual systems are in the same pure state and that is all the information we have available. The laws of nature allow that different outcomes are possible even when we perform the same measurement in two identically prepared systems.

In this text, we present a number of attempts to find objective criteria which allow us to decide between these two options. We will see that, under some very reasonable circumstances, there is no way out but to accept option II.

Before we enter the specific details of the proofs of the impossibility of option I, let us think about why this option seems so logical to our classical minds, modelled by our daily experience with macroscopic systems. The necessity of the use of probabilities in the description of an experiment naturally arises from the incompleteness of our knowledge about the parameters involved in it. According to our classical viewpoint, if we knew everything about our experiment, two repetitions of the same procedure with exactly the same value for every possible parameter involved would provide the same result at the end. It is reasonable to imagine that two replicas of the same object will remain identical if they are subjected to exactly the same process. If this is not the case, we would probably not call them identical at all.

This is not what happens with quantum systems. Even if all elements of an ensemble are in the same pure quantum state, quantum theory predicts that for most measurements with exclusive outcomes, more than one of them have positive probability. Consequently, according to the argument in the previous paragraph, the elements of the ensemble could not be identical and hence the state assigned to this preparation by quantum theory could not be everything: there would be more parameters we must use in the description of these systems in order to get definite outcomes for all measurements. These unknown parameters may have different values in our ensemble, and the probabilistic behaviour is due to our lack of knowledge on these "hidden variables."

This line of thought led many physicists to believe that quantum theory might be wrong, or at least, incomplete [EPR35]. Since quantum theory is consistent with the statistics obtained from every experiment made so far, yet unable to foresee the precise outcome for each run of each experiment, we have no empirical reason for discarding it. Hence, the best shot is to suppose the possibility of completing quantum theory, adding extra variables to the description of pure states, in a way that with all this information (of pure quantum state plus extra variables) we would be able to predict with certainty the outcome of all measurements and in a way that when averaging over these extra variables we would recover the quantum predictions.

A good example in which a similar argument applies is classical thermodynamics, which states physical laws involving macroscopic aspects of matter, such as pressure, volume and temperature. These laws do not provide complete information about the systems studied, since they appear when we average over a large number of atoms and we do not take into account the individual parameters such as position and velocity of each atom. Although very useful for many applications, classical thermodynamics does not explain phenomena such as Brownian motion, which require a more complete treatment, provided by statistical physics.

It happens that under the assumption of *noncontextuality*, completions of quantum theory are generally not possible. This result is known as the *Bell-Kochen-Specker theorem*. The noncontextuality hypothesis states that the value assigned to a measurement in the completion cannot depend on other compatible measurements performed jointly.

The pioneer proof of this result was provided by Kochen and Specker [KS67]. It is based on a set of 117 measurements with possible outcomes 0 or 1 in a three-dimensional quantum system. With this set, we can define a graph whose vertices correspond to the measurements[1] and whose edges connect measurements that are exclusive. The noncontextuality hypothesis and quantum theory in dimension three imply that we cannot assign the value 1 to two connected vertices and that in a triangle we must have at least one vertex assigned the value 1. This set of measurements is constructed in such a way that these rules cannot be both satisfied.

The assumption of noncontextuality in Kochen and Specker's proof was so natural that it was only pointed out later by Bell [Bel66]. Many other proofs using the same idea have been provided, using sets with a smaller number of measurements. They have an important common feature: they are all *state-independent*. This means that if we choose the set of projective measurements as in any of these proofs, the assignment of definite values for them cannot reproduce the statistics given by any quantum state when we average over all possible values of the extra variables. The reader interested in such proofs may find a number of examples in Appendix A.

It is possible to provide simpler state-dependent proofs of the impossibility of completions of quantum theory. The idea behind this kind of proof is to show that no such completion can reproduce the statistics of a *specific* set of measurements in a *specific* state of the corresponding system. Some of these proofs use a very small number of measurements and hence are much simpler than the state-independent ones.

One of the most common ways to provide a state-dependent proof of the Kochen-Specker theorem is using the so-called *noncontextuality inequalities*. They are linear inequalities involving the probabilities of outcomes of joint measurements that must be obeyed by any noncontextual completion of quantum theory. Some of these inequalities can be violated with quantum systems with a particular choice of state and measurements, which establishes the phenomenon of quantum contextuality.

---

[1] In fact, three out of these 117 measurements count twice, giving rise to a 120-vertex graph (see Fig. A.4).

One advantage of the impossibility proofs using noncontextuality inequalities is that many of them use a small number of measurements, which may make them much more suitable for experimental implementations. The experimental verification of quantum violations was already performed for a number of inequalities, specially in the particular case of Bell scenarios [Asp75, HLB⁺06, KZG⁺09, HLZ⁺09, LLS11, BCdA⁺14, HBD⁺15, GVW⁺15, SMSC⁺15, HKB⁺16, CAC⁺16, LMZNCAH18].

The quest for proofs of the impossibility of noncontextual completions of quantum theory led to the development of a beautiful mathematical framework, after the recognition that many elements of graph theory and combinatorics appear naturally in the study of contextuality. In this book, we present the main elements of this fruitful connection between contextuality and graphs. We start with a brief presentation of the abstract mathematical description of states and measurements, called a *probabilistic model*, and define precisely what we mean by *classical probabilistic models* and *quantum probabilistic models*. We discuss completions of probabilistic models and the assumption of noncontextuality, the basic elements of the impossibility proofs presented in the next chapters.

## 1.1  States and Measurements

In this section, we discuss briefly the suitable mathematical structure used in the description of experiments carried in a hypothetical physical system. Throughout the book, we adopt an operational view of *preparations* and *measurements*, general enough to allow classical probability theory, quantum theory and some other generalisations of probability theory. This approach is similar to those presented in Refs. [Bar07, LSW11, BW16, Ama14], where the curious reader can find a deeper discussion of the subject.

Our first assumption is about the nature of the experiments that can be performed in this system. We assume that there are two kinds of interventions available: *preparations* and *operations*. Another important requirement is that these experiments be reproducible: every preparation and every operation can be performed as many times as needed and we can use several repetitions of a given procedure to count relative frequencies. For each operation, there may be several different outcomes, each occurring with a well-defined probability for a given preparation. Preparations can be compared through their statistics in relation to the given operations, and these statistics define a *state*.

**Definition 1.1** Two preparations are equivalent if they give the same probability distribution for all available operations. The equivalence class of preparations is called a *state*.

We assume that the set of all states, called the *state space* of the system, is a convex set, whose extremal points are called *pure states*.

**Definition 1.2** Operations with more than one outcome are called *measurements*.

The mathematical description of a physical system must provide a set of objects to represent the states of the system and a set of objects to represent the operations and measurements. It must also provide a rule to calculate the probabilities of the possible outcomes of a measurement given the state. Such a description is called a *probabilistic model*. A *probability theory* is a collection of probabilistic models.

## 1.1.1 Repeatability and Compatibility

When we talk about contextuality, it is usual to restrict the discussion to the special class of *outcome-repeatable measurements*. Outcome-repeatability is a characteristic of a measurement that restricts the possible outcomes if this measurement is applied sequentially to the *same system*. Notice that this is different from two independent realisations of the same experiment, where two copies of the system are prepared in the same state and are then subjected to the same measurement.

**Definition 1.3** A measurement $j$ has *repeatable outcomes* if every time this measurement is performed in a system and an outcome $k$ is obtained, a subsequent measurement of $j$ in the same system gives outcome $k$ with probability one.

All measurements are assumed to be outcome-repeatable from now on. We denote by $p_j(k)$ the probability of obtaining outcome $k$ given that the measurement $j$ was performed. This naturally depends on the preparation, but we will not make this dependence explicit on the notation.

One of the implications of more general probability theories is that there may not be a well-defined joint probability distribution for the results of *all* measurements in a given set. When this global probability distribution exists for all states, we say that the measurements are *compatible*. This should not be new for the reader familiar with quantum theory, where noncompatibility is the rule, not the exception.

**Definition 1.4** A set of outcome-repeatable measurements $\{j_1, \ldots, j_n\}$ is *compatible* if there is another measurement $j$ with outcomes $\{1, \ldots, m\}$ and functions $f_1, \ldots, f_n$ such that the possible outcomes of each $j_s$ are $f_s(\{1, \ldots, m\})$ and

$$p_{j_s}(l) = \sum_{k \in f_s^{-1}(l)} p_j(k). \tag{1.1}$$

The measurement $j$ is called a *refinement* of each $j_i$, and each $j_i$ is called a *coarse graining* of $j$.

If the measurements $\{j_1, \ldots, j_n\}$ are compatible, the probability of a set of outcomes $l_1, \ldots, l_n$ given that $j_1, \ldots, j_n$ where jointly measured is well defined and it is equal to the probability of outcomes $\bigcap_s f_s^{-1}(l_s)$ for measurement $j$.

The notion of compatibility is essential in quantum theory. It is connected to the idea of "measurements that can be performed jointly." If a set of measurements is compatible, they can be measured jointly on the same individual system without disturbing the results of each other. By construction, it is equivalent to measure all $j_k$, in any order, or to measure $j$ in Definition 1.4 and then use functions $f_k$ to find out the outcomes of each $j_k$.

### 1.1.2  Classical Probability Theory

The axiomatic system we use to describe classical probability theory was introduced by the soviet mathematician Andrey Kolmogorov in the 1930s [SW95, GS01, Jam81]. Although classical probabilistic models can be used to describe a large variety of random phenomena, they are not suitable to describe the behaviour of quantum systems.

All axioms in classical probability models seem very natural and it is surprising that nature does not always behave in this way. These axioms imply a number of singular properties that make this kind of theory different from any other in the framework treated in this book. In this sense, classical theory emerges as a very special exception.

In this book, we assume that measurements have a finite number of outcomes and we will typically work with a finite number of measurements.

A classical model depends on three objects: a sample space $\Omega$, a $\sigma$-algebra $\Sigma$ in $\Omega$ and the set of probability measures $\mu$ in the measurable space $(\Omega, \Sigma)$.

**Definition 1.5** A *classical probabilistic model* over a measurable space $(\Omega, \Sigma)$ is given by a set of states and measurements in which every state is a probability measure $\mu$ in $(\Omega, \Sigma)$ and every measurement corresponds to a partition $\{A_i\}$ of the sample space $\Omega$ into measurable sets $A_i \in \Sigma$. Given a measurement $\{A_i\}$ and a state $\mu$, the probability of outcome $i$ is

$$p(i) = \mu(A_i). \tag{1.2}$$

**Definition 1.6** A *classical probability theory* is one in which all models are classical.

For measurements with repeatable outcomes in classical probability theory, there are no incompatible measurements. This is quite easy to prove: it is a consequence of the fact that a finite intersection of sets in a $\sigma$-algebra is also an element of the $\sigma$-algebra.

**Theorem 1.1** *In a classical probabilistic model, all measurements with repeatable outcomes are compatible.*

*Proof* Let $\{j\}$ be a finite set of measurements in a classical model, with $j = \left\{A_i^j\right\}$ a measurement with a finite number of outcomes for each $j = 1, \ldots, n$. For each list of outputs $(i_j)$, define $R_{(i_j)} = \left\{\cap_{j=1}^n A_{i_j}^j\right\}$. Since each element in the set $\left\{R_{(i_j)}\right\}$

is a measurable set, it defines a measurement that is a common refinement for all measurements $j$.                                                                      □

When the sample space is a finite set, the maximal measurement for which every set $A_i$ contains only one element is a common refinement for all other measurements.

## 1.1.3  Quantum Probability Theory

Quantum theory is, at the same time, the first physical theory where the probabilistic character is considered intrinsic, and the first physical theory which does not fit into classical probabilistic models under reasonable assumptions.

**Definition 1.7** A *quantum probabilistic model* over a complex Hilbert space $\mathfrak{H}$ is a set of states and measurements in which every state is a positive-semidefinite operator $\rho$ acting on $\mathfrak{H}$ such that $\mathrm{Tr}\,(\rho) \leq 1$ and every measurement[2] is a partition of the identity operator

$$I = \sum_i P_i \tag{1.3}$$

in terms of orthogonal projectors $\{P_i\}$. Given a measurement $\{P_i\}$ and a state $\rho$, the probability of outcome $i$ is

$$p(i) = \mathrm{Tr}\,(P_i \rho). \tag{1.4}$$

**Definition 1.8** A measurement $\{P_i\}$ is called *complete* if each $P_i$ is a projector into a one-dimensional subspace.

Compatibility of two outcome-repeatable measurements can be easily decided in quantum models from the projectors defining the measurements.

**Theorem 1.2** *For two projective measurements* $\{P_1, \ldots, P_n\}$ *and* $\{Q_1, \ldots, Q_m\}$, *the following are equivalent:*

1. $\left[P_i, Q_j\right] = 0$ *for every* $1 \leq i \leq n$ *and* $1 \leq j \leq m$;
2. *Both measurements are coarse grainings of the same complete projective measurement;*
3. *The measurements are compatible.*

---

[2]To be precise, this is the definition of a *projective measurement*, which is not the most general measurement one can perform in a quantum system. However, repeatability suggests the restriction to projective measurements when dealing with contextuality. We will restrict our definition to this special class of measurements, and the word *measurement* in this book will always mean *projective measurement* for quantum systems. To define the most general measurement in quantum theory, one needs the notion of Generalised Measurement, related to the notion of POVM (positive operator valued measure), and the interested reader can find the definition in Ref. [NC00].

*Proof* Equivalence between items 2 and 3 is a direct consequence of Definition 1.4. To prove that these conditions are equivalent to 1, we notice that both measurements are coarse grainings of the same complete projective measurement iff all $P_i$ and $Q_j$ are simultaneously diagonalisable, and hence, iff they commute.                                                           □

Classical and quantum probability theories are just two examples among many others. In this book, we discuss the interesting differences between the two, but we are also interested in the differences between them and other general probabilistic theories. In particular, we are concerned with the features that distinguish them from this landscape of possible theories. Classical models have many unique properties that distinguishes them from all other probabilistic models (see Refs. [Bar07, LSW11, BW16, Ama14]), but the unique properties of quantum models that force this formalism upon our description of quantum systems are yet unknown. A promising candidate, the *Exclusivity Principle*, and its consequences are discussed in Chap. 4.

## 1.2   Completing a Probabilistic Model

The idea that quantum theory is intrinsically probabilistic puzzled those who worked on its construction since the early 1920s. The fact that it can only provide probabilities, no matter how well one knows the system, is in contrast with all the belief built by the previous experiences with classical systems and the corresponding classical description. Hence, many conjectured the possibility of a more complete description that would associate to each quantum probabilistic model a classical probabilistic model, possibly with a huge sample space, such that the predictions of the quantum model would be recovered when averaged over adequate sets of states in this classical model. As we mentioned before, the aim of this book is to describe many proofs that this is not possible and some of the remarkable consequences that come from this impossibility.

To this end, we do not need to consider a very complicated quantum probabilistic model. It suffices to consider a single state and a set of measurements chosen in some special way. As we will see, the notion of *incompatible* measurements plays a crucial role in this choice. In the next chapters, we show several examples of states and measurements that will do the job, as well as strategies to obtain them, and some unexpected relations to graph theory. Without more delay, let us introduce some important definitions.

**Definition 1.9** Let $X$ be the set of measurements in a probabilistic model. A set of compatible measurements $\{j_1, j_2, \ldots, j_m\} \subset X$ is called a *context*. We say that this context is *maximal* if there is no other set of compatible measurements in $X$ that contains it properly.

**Definition 1.10** Let $S$ be the set of pure states and $X$ be the set of measurements in a probabilistic model P. We say that a probabilistic model P′ is a *completion* for P if the measurements of P' are in one-to-one correspondence with the measurements in $X$, the pure states of P' are in one-to-one correspondence with $\Lambda \times S$ for some set $\Lambda$ and for each $\rho \in S$ there is a probability distribution $p(\lambda)$ over $\Lambda$ such that

$$p_C^\rho(i_1 \ldots i_m) = \sum_{\lambda \in \Lambda} p(\lambda) \prod_{k=1}^m p_{j_k}^{(\lambda, \rho, C)}(i_k) \qquad (1.5)$$

for each context $C = \{j_1 \ldots j_m\}$, where $p_C^\rho(i_1 \ldots i_m)$ is the probability of outcomes $i_1 \ldots i_m$ for a joint measurement of $j_1 \ldots j_m$ in state $\rho$ and $p_{j_k}^{(\lambda, \rho, C)}(i_k) \in \{0, 1\}$ determines with certainty which outcome $i_k$ is obtained for measurement $j_k$ in state $(\lambda, \rho) \in \Lambda \times S$.

Intuitively, a completion for a probabilistic model is a new model in which the states contain the information of the states in $S$ plus the information contained in an extra variable $\lambda$. With the complete information of $\rho$ and $\lambda$, it is possible to predict with certainty the outcome of each measurement in $X$, condition guaranteed by the constraint $p_{j_k}^{(\lambda, \rho, C)}(i_k) \in \{0, 1\}$. Equation (1.5) then implies that the probabilities in the original probabilistic model are recovered when we average over $\lambda$. If a completion of this sort is possible, we essentially recover the classical intuition: probabilities become necessary only because, for some unknown reason, we lack the knowledge of the extra variable[3] $\lambda$. If we could design a clever way to learn $\lambda$, we would be able to predict with certainty the results of all measurements in $X$.

### *1.2.1   The Assumption of Noncontextuality*

Let $C_1 = \{j_1, j_2, \ldots, j_m\}$ and $C_2 = \{j_1, j_2', \ldots, j_n'\}$ be two contexts containing $j_1$ in a probabilistic model P. Measurements $j_k$ and $j_l'$ are not necessarily compatible for $k > 1$. Suppose that P' is a completion for P. Each pure state $(\lambda, \rho)$ in P' assigns a string of definite values to both $\{j_1, j_2, \ldots, j_m\}$ and $\{j_1, j_2', \ldots, j_n'\}$. The values assigned to the measurements in context $C_1$ are determined by the functions $p^{(\lambda, \rho, C_1)}$ and the values assigned to the measurements in $C_2$ are determined by the functions $p^{(\lambda, \rho, C_2)}$. Notice that the definition of completion does not necessarily imply that

$$p_{j_1}^{(\lambda, \rho, C_1)}(i) = p_{j_1}^{(\lambda, \rho, C_2)}(i), \qquad (1.6)$$

---

[3]The reader familiar with the notion of *hidden-variable model* will notice that it is equivalent to our notion of completion. This term comes from the idea of existence of a deeper theory describing some underlying reality. However, it is important to say that such original motivation is superfluous. A completion is simply a classical probabilistic model extending the given behaviour, no matter how one interprets probabilities.

but this is a natural constraint if we aim to reconcile the probabilistic model P with classical probability theory. Then, we demand that the value assigned to $j_1$ by a pure state in P′ be independent of the context in which it appears: if the outcome of $j_1$ according to the completion is $i_1$ when a joint measurement of $\{j_1, j_2, \ldots, j_m\}$ is performed, the same outcome $i_1$ must be assigned to $j_1$ by the completion if we jointly measure $\{j_1, j_2', \ldots, j_n'\}$.

**Definition 1.11** We say that a completion of a probabilistic model P is *noncontextual* if Eq. (1.6) is satisfied for all measurements $j_1$ and contexts $C_1, C_2 \ni j_1$.

In a noncontextual completion, the value associated by the model to any measurement $j_1$ is context independent, i.e., it is the same regardless of which other mutually compatible measurements are jointly performed or not. This observation was first pointed out by Bell [Bel66], who argued that there is no *a priori* reason to require noncontextuality from a completion. Suppose we perform the measurement of $j_1$ and together we may choose to measure either $\{j_2, \ldots, j_m\}$ or $\{j_2', \ldots, j_n'\}$, both compatible with $j_1$ but not to one another. These different possibilities may require completely different experimental arrangements, and hence to demand that the values associated to $j_1$ be the same for both contexts cannot be physically justified. The outcome of a measurement might depend not only on the state of the system but also on the apparatus used to measure it.

Although the measurement process and the interaction between system and apparatus are important issues in quantum theory, this is not the problem here, since we could include all variables of the apparatus in the model, and apply the same reasoning again. The point that makes the noncontextuality assumption plausible is that there is no need to perform the compatible measurements simultaneously. Suppose we measure $j_1$ and then we choose what else we are going to measure, $\{j_2, \ldots, j_m\}$ or $\{j_2', \ldots, j_n'\}$, or even if we are not measuring anything else. The completion should predict the outcome of $j_1$, but if this model is contextual this value would depend on a measurement that will be performed in the future [Whe78] or, even worst, on a decision to measure or not, yet to be made!

Another way to naturally enforce the noncontextuality assumption is to design the experiment in such a way that the choice of $\{j_2, \ldots, j_m\}$ or $\{j_2', \ldots, j_n'\}$ is made in a different region of the space in a time interval that forbids any signal to be sent from one region to the other. Since no signal was received, the choice of what is going to be measured in one part cannot depend on what happens in the other, demanding the completion to be noncontextual. In this situation, we say that the completion is local and the noncontextuality assumption is usually referred to as the *locality assumption*.

## 1.3 Outline of the Book

In this text, we discuss in detail the assumption of noncontextuality and the impossibility of completions of general probabilistic models, in particular quantum probabilistic models. We focus mainly on two different approaches, both connected with the successful graph-theoretical formulation of quantum contextuality, which supplies new tools to understand the differences between quantum and classical theories and also the differences between quantum theory and more general theories [Cab13b, Yan13, ATC14].

In Chap. 2, we present the compatibility-hypergraph approach, where the main concerns are the compatibility relations of a finite set of measurements. We will see in detail the different convex sets that arise when we compute the probabilities of the joint measurement of each context using general probability theories that satisfy the *nondisturbing condition*, which include classical and quantum probability theories as special cases. With this characterisation, noncontextuality inequalities arise naturally from the geometric description of the classical set. We analyse the relation between noncontextuality inequalities and the *exclusivity graph* of the contextuality scenario and how the classical and quantum bounds to these inequalities are related to graph invariants of this graph. We then look for the special case in which every context consists of at most two binary measurements, in which both nondisturbing and noncontextual sets can be equivalently described in different ways that lead to familiar polytopes from graph theory. We finish this chapter with a brief introduction to the sheaf theoretical aspects of contextuality, which provide a direct and unified characterisation of both contextuality and non-locality, along with different new tools, insights and results.

In Chap. 3, we present the exclusivity-graph approach to contextuality, where the main concerns are the exclusivity relations of a finite set of measurement events. We will see that in this approach the convex sets obtained when the probabilities are computed with classical, quantum and general probability theories satisfying the so-called *Exclusivity Principle* are related to important convex sets of graph theory. Noncontextuality inequalities also arise from the geometric description of the classical set and the classical and quantum bounds are related to graph invariants of the exclusivity graph. This is also the case for the bound obtained with probability theories satisfying the exclusivity principle. We then discuss two refinements of the exclusivity-graph approach. First, we present the coloured-graph approach to nonlocality, where each party exclusivities in a Bell scenario are encoded in a different colour for the edges of the exclusivity graph. Second, we present the exclusivity-hypergraph approach, where the idea is not only to take into account the exclusivity relations among the measurement events but also to include the information of which measurement gave rise to each specific exclusivity.

After introducing the graph approaches to contextuality, we discuss how they lead to important developments in foundations of quantum physics. One of the most thought-provoking scientific challenges in recent times is deriving quantum theory from first principles. The starting point is assuming general probabilistic theories allowing for correlations that go beyond those that arise in quantum theory, and the goal is to find principles that pick out quantum theory from this landscape of possible theories. The exclusivity principle was proposed as a possible answer and many results have been found so far. A detailed review of these results can be found in Chap. 4.

In Appendix A, we go back in time to present a brief historical development of the proof of the Bell-Kochen-Specker theorem [Bel66, KS67]. We start from von Neumann's challenge to the idea of completing quantum theory, which appears in his pioneer rigorous mathematical formulation for quantum theory [vN55]. One of his assumptions was too strong, but his result was a landmark in foundations of physics, since it opened the door for a series of papers proving the impossibility of this kind of completion. We then present Gleason's lemma [Gle57], that can be used to discard the possibility of noncontextual completions of Quantum Theory. Although the Bell-Kochen-Specker theorem follows from Gleason's lemma, this fact was noticed only after it was proved by other means by Kochen and Specker. The advantage of Kochen and Specker's proof is that, contrary to Gleason's lemma, it uses only a finite number of projectors. The Kochen and Specker's original proof is presented, along with other simple additive and multiplicative proofs. We finish this chapter presenting a contextual completion of a Quantum Probabilistic Model, showing that the noncontextuality hypothesis is indeed crucial to discard the possibility of completions.

# Chapter 2
# Contextuality: The Compatibility-Hypergraph Approach

Interesting discrepancies between probability theories emerge when we look at the statistics of a finite set of measurements divided among several contexts with non-empty intersection. In this chapter we will see that one of these discrepancies is related to the unexpected phenomenon of *contextuality*. This phenomenon is deeply connected to *incompatibility of measurements* and hence only arises in probability theories other than classical, since all measurement are compatible in classical models. This makes classical probability theory a very unique exception.[1]

The statistics of a finite set of measurements with given compatibility relations gives rise to different convex sets depending on the probability theory used to describe states and measurements. For classical probability theory, this set is a polytope [Roc70], hence defined by a finite set of linear inequalities. Membership in this set is equivalent to the existence of a noncontextual completion reproducing the given statistics. For this reason, this polytope is known as the *noncontextual set* and the nontrivial linear inequalities that define it are called *noncontextuality inequalities*. Statistics that are outside this set are named *contextual*. The set obtained with quantum theory is in general strictly larger than the noncontextual set, a result that demonstrates the phenomenon of *quantum contextuality*.

Quantum contextuality can already be witnessed with very simple sets of measurements. One of the most elementary noncontextuality inequalities which is not always satisfied in quantum theory is the KCBS inequality, named after the pioneer work of Klyachko, Can, Binicioğlu, and Shumovsky [KCBS08], which involves only five measurements. This inequality can be generalised to similar

---

[1]It is interesting and intriguing the fact that, in the space of possible generalised probability theories, classical theory is defined by a simple property: all measurements are compatible. We lack such characterisation for quantum theory. This is the goal of the so-called search for "principles" which lead to quantum theory, a rich line of research usually referred to as *reconstructions of quantum theory* [Har01, Har11, DvB11, MM11, CDP11, Koc15, AG16].

© The Author(s), under exclusive licence to Springer Nature Switzerland AG 2018
B. Amaral, M. Terra Cunha, *On Graph Approaches to Contextuality and their Role in Quantum Theory*, SpringerBriefs in Mathematics,
https://doi.org/10.1007/978-3-319-93827-1_2

inequalities involving $n$ measurements, the $n$-cycle inequalities, which can be used to demonstrate quantum contextuality for $n > 3$. Besides the importance of these inequalities due to their simplicity, crucial for experimental implementations, the sets of measurements that lead to them are the building blocks for contextuality, since any set of measurements exhibiting contextuality must contain one of these blocks as a subset.

The compatibility relations among the set of measurements can be encoded in the *compatibility hypergraph*, whose vertices are the measurements in the scenario and the hyperegdes are the contexts. The structure of this hypergraph on its own already affects the geometry of the noncontextual, quantum, and nondisturbing sets, as described in Sect. 2.2, which provides a beautiful combination of contextuality and graph theory. This combination is even stronger if each context consists of at most two binary measurements, in which both nondisturbing and noncontextual sets can be equivalently described in different ways that lead to familiar polytopes from graph theory.

From the compatibility hypergraph, we can construct the *exclusivity graph* of the scenario, whose vertices are measurement events and edges connect *exclusive events*, as described in Sect. 2.4. The characterisation of the maximum value of some linear functions defined on these convex sets, which include the noncontextuality inequalities, is related to graph invariants of the exclusivity graph. All these important results are just a taste of the power of the graph approach to contextuality, first presented by Cabello, Severini and Winter in Refs. [CSW10, CSW14], and later developed by many other authors [AFLS15, RDLT$^+$14].

We finish this chapter with a brief introduction to the formal mathematical description of contextuality in terms of categories and sheaf theory. This approach provides a direct and unified characterisation of both contextuality and non-locality, along with different new tools, insights and results.

## 2.1  Compatibility Scenarios

**Definition 2.1**  A *compatibility scenario*[2] is defined by a triple $\Upsilon := (X, \mathscr{C}, O)$, where $O$ is a finite set, $X$ is a finite set of random variables in $(O, \mathscr{P}(O))$ and $\mathscr{C}$ is a family of subsets of $X$ such that

1. $\cup_{C \in \mathscr{C}} C = X$;
2. $C, C' \in \mathscr{C}$ and $C \subseteq C'$ implies $C = C'$.

The elements $C \in \mathscr{C}$ are called (maximal) *contexts* and the set $\mathscr{C}$ is called the *compatibility cover* of the scenario.

The physical interpretation is that the random variables in $X$ represent measurements with possible outcomes $O$ in a probabilistic model and the sets in $\mathscr{C}$ encode

---

[2]We will often use the word *scenario* instead of *compatibility scenario*.

the compatibility relations among the elements of $X$, that is, each set $C \in \mathscr{C}$ consists of a maximal set of compatible measurements. This implies that the elements in a context $C$ represent measurements that can be jointly performed. Equivalently, the compatibility relations among the elements of $X$ can be represented with the help of a hypergraph.

**Definition 2.2** The *compatibility hypergraph* of a scenario $(X, \mathscr{C}, O)$ is a hypergraph $\mathscr{H} = (X, \mathscr{C})$ whose vertices are the measurements in $X$ and the hyperedges are the contexts $C \in \mathscr{C}$. The *compatibility graph* of the scenario is the 2-skeleton of $\mathscr{H}$, that is, the graph $\mathscr{G}$ with the same vertices of $\mathscr{H}$ and edges connecting $i, j \in X$ iff there is a context $C \in \mathscr{C}$ such that $i, j \in C$.

In some probabilistic theories, the fact that a set of measurements is pairwise compatible does not imply that this set is jointly compatible. For example, if we have a set of three measurements $X = \{M_1, M_2, M_3\}$ with contexts $\{M_1, M_2\}$, $\{M_2, M_3\}$ and $\{M_1, M_3\}$, the set $\{M_1, M_2, M_3\}$ is not necessarily a context. Notice, however, that the compatibility graph is the same in both situations. Hence, in general, the compatibility hypergraph is more subtle than its 2-skeleton. On the other hand, if only quantum contextuality with projective measurements is to be discussed, then the compatibility graph is enough, since pairwise compatible projectors are jointly compatible. The maximal contexts in this case correspond to cliques of the compatibility graph.

For a given context $C \subset \mathscr{C}$, consider the set of possible outcomes for a joint measurement of the elements of $C$. This set is the Cartesian product of $|C|$ copies of $O$ and will be denoted by $O^C$. It can be identified with the set of functions

$$s : C \longrightarrow O. \tag{2.1}$$

When the measurements in $C$ are jointly performed, a set of outcomes in $O^C$ will be observed. This individual run of the experiment will be called a *measurement event*.

**Definition 2.3** A *behaviour* B for the scenario $(X, \mathscr{C}, O)$ is a family of probability distributions over $O^C$, one for each context $C \in \mathscr{C}$, that is,

$$\mathrm{B} = \left\{ p_C : O^C \to [0, 1] \,\middle|\, \sum_{s \in O^C} p_C(s) = 1, C \in \mathscr{C} \right\}. \tag{2.2}$$

For each $C$, $p_C(s)$ gives the probability of obtaining outcomes $s$ in a joint measurement of the elements of $C$. It will be convenient to associate each behaviour to a vector $P_\mathrm{B} \in \mathbb{R}^d$, with

$$d = \sum_{C \in \mathscr{C}} \left| O^C \right|. \tag{2.3}$$

Denoting $\mathscr{C} = \{C_1, C_2, \ldots, C_n\}$ and $O^{C_i} = \{s_i^1, s_i^2, \ldots, s_i^{m_i}\}$ for each $C_i$, we define

$$P_B = \left[ p_{C_1}\left(s_1^1\right) p_{C_1}\left(s_1^2\right) \cdots p_{C_1}\left(s_1^{m_1}\right) \cdots p_{C_n}\left(s_n^1\right) p_{C_n}\left(s_n^2\right) \cdots p_{C_n}\left(s_n^{m_n}\right) \right].$$

(2.4)

This association is discussed in more detail in Ref. [AQB$^+$13].

For a given compatibility cover, the set of possible behaviours is a polytope with

$$\prod_{C \in \mathscr{C}} \left| O^C \right|$$

(2.5)

vertices. Each vertex corresponds to probability one for one of the outcomes $s \in O^C$ for each context $C \in \mathscr{C}$. All other behaviours are convex combinations of these vertices.

Let $C = \{M_1, \ldots, M_m\}$ be a context in $\mathscr{C}$. Each element of $O^C$ is a string $s = (a_1, \ldots, a_m)$ with $m$ elements of $O$. For each $U \subset C$, there is a natural restriction

$$r_U^C : O^C \to O^U$$

$$s = (a_i)_{M_i \in C} \mapsto s|_U = (a_i)_{M_i \in U}.$$

(2.6)

This operation corresponds to dropping the elements in the string $s$ that do not correspond to measurements in $U$.

Given a probability distribution in $C \in \mathscr{C}$, we can also naturally define marginal distributions for each $U \subset C$:

$$p_U^C : O^U \to [0, 1]$$

$$p_U^C(s) = \sum_{s' \in O^C; r_U^C(s') = s} p_C(s').$$

(2.7)

The superscript $C$ in $p_U^C$ is necessary because, without further restrictions, the marginals may depend on the context $C$.

*Example 2.1* Consider the situation where

$$X = \{M_1, M_2, M_3\} \text{ and } \mathscr{C} = \{C_1 = \{M_1, M_2\}, C_2 = \{M_2, M_3\}\},$$

(2.8)

each measurement with two possible outcomes $\pm 1$. The extreme distribution with $p_{C_1}(1, 1) = 1$ and $p_{C_2}(-1, -1) = 1$ gives the marginals $p_{M_2}^{C_1}(1) = 1$ and $p_{M_2}^{C_2}(1) = 0$.

Example 2.1 shows a case of trivial contextuality: the answer for $M_2$ is determined by the context. In this book, we focus on nontrivial cases: we require that if two contexts $C_1$ and $C_2$ overlap, the marginals generated by $p_{C_1}$ and $p_{C_2}$ in the intersection be the same.

**Definition 2.4**  Given a scenario $\Upsilon$, the *nondisturbance set* $\mathscr{X}(\Upsilon)$ is the set of behaviours such that for any two intersecting contexts $C$ and $C'$ the consistency relation $p^C_{C \cap C'} = p^{C'}_{C \cap C'}$ holds.

For each given scenario, the nondisturbance set is a polytope, since it is defined by a finite number of linear inequalities and equalities: the inequalities imposed by the fact that its elements represent probabilities and the equalities imposed by Definition 2.4. From now on, unless explicitly stated otherwise, we assume that all behaviours are nondisturbing.

After imposing conditions on the restriction of the probability distributions, we ask now if it is possible to do the opposite: extend the distributions $p_C$ to larger sets containing $C$ in a *consistent* way. The naive ultimate goal would be to define a distribution on the set $O^X$, which assigns outcomes to every measurement, in a way that the restrictions yield the probabilities specified by the behaviour on all contexts in $\mathscr{C}$. A more subtle and adequate question is to decide when it is possible to achieve this goal. This question was first studied by Fine in Ref. [Fin82], for the restricted case of Bell scenarios, and generalised by Brandenburger and Abramsky in Ref. [AB11]. As it happens in many branches of mathematics, the notion of contextuality is deeply connected to the possibility of extending elements of $O^C$ to *global sections* in $O^X$.

**Definition 2.5**  A *global section* for $X$ is a probability distribution $p_X : O^X \to [0, 1]$. A *global section for a behaviour* $B \in \mathscr{X}(\Upsilon)$ is a global section for $X$ such that the restriction of $p_X$ to each context $C \in \mathscr{C}$ is equal to $p_C$. The behaviours with global section are called *noncontextual*.

Each $s \in O^X$, seen as function from $X$ to $O$, gives definite outcomes for a joint measurement of $X$: the value assigned by $s$ to a measurement $M \in X$ is $s(M)$. The global section $p_X$ corresponds to a convex combination of such deterministic instances. Hence, a global section for a behaviour $B$ corresponds exactly to the existence of such a convex combination which marginalised yields the probabilities determined by the behaviour. This is deeply connected to the existence of a noncontextual completion reproducing the statistics of $B$. In fact, if there is a global section for $B$, we can construct the completion in the following way: as the set $\Lambda$ in the completion we use an element of the classical probability space $O^X$. Then, the global section $p_X$ for $B$ provides a probability distribution in the set $\Lambda$ with the property that if we average over all values of $\lambda$ according to this function, we recover the predictions of the behaviour $B$. A proof of the converse can be found in section 8 of Ref. [AB11], and this gives:

**Theorem 2.1 (Fine, Brandenburger and Abramsky, 2011)** *A behaviour* $B \in \mathscr{X}(\Upsilon)$ *has a global section if and only if there is a noncontextual completion recovering its statistics.*

Some behaviours do not admit global sections. They are called *contextual*.

*Example 2.2 (Contextual Nondisturbing Behaviour)* Consider the scenario $(X, \mathscr{C}, O)$, where

$$X = \{M_1, M_2, M_3\}, \quad \mathscr{C} = \{\{M_1, M_2\}, \{M_2, M_3\}, \{M_1, M_3\}\} \quad \text{and} \quad O = \{-1, 1\}.$$
(2.9)

The behaviour given by

|          | $(1, 1)$      | $(1, -1)$     | $(-1, 1)$     | $(-1, -1)$    |
|----------|---------------|---------------|---------------|---------------|
| $M_1 M_2$ | $\frac{1}{2}$ | $0$           | $0$           | $\frac{1}{2}$ |
| $M_2 M_3$ | $\frac{1}{2}$ | $0$           | $0$           | $\frac{1}{2}$ |
| $M_1 M_3$ | $0$           | $\frac{1}{2}$ | $\frac{1}{2}$ | $0$           |

where each line specifies a context, $C$, each column defines a set of outcomes, $s$, and the corresponding value on the table is the probability of obtaining $s$ when measuring $C$, is a nondisturbing behaviour, but it does not have a global section.

### 2.1.1 Bell Scenarios

When the system under consideration is composed of $n$ different spatially separated subsystems, the set $X$ can be divided into various distinct subsets $X_1, X_2, \ldots, X_n$, where $X_i$ is the set of measurements available for party $i$. In this case, compatibility is guaranteed by the spatial separation among the parties, and all contexts are of the form

$$C = \{M_1, M_2, \ldots, M_n\}, \quad M_i \in X_i.$$
(2.10)

Scenarios with these extra restrictions are called *Bell scenarios*. The particular case in which each party has $m$ measurements available, each measurement with $o$ possible outcomes, is denoted by $(n, m, o)$.

The vertices of the compatibility hypergraph of a Bell scenario can be split into the $n$ disjoint subsets $X_i$ and each hyperedge contains one, and only one, element of each $X_i$. Hence, the compatibility graph is the complete $n$-partite graph $G(X_1, \ldots, X_n)$.

The nondisturbance condition in this case receives a natural and important interpretation: no message can be sent from one part to another by using a nondisturbing behaviour [BJ87]! For this reason, for Bell scenarios it is known as the nosignalling condition. In Bell experiments, one usually invokes the notion that no signal can travel faster than the speed of light in order to assume this condition for any possible behaviour.

A noncontextual completion for a behaviour in such a scenario is a convex combination of assignments of definite values to each measurement that are independent of the measurements performed by any other party. Hence, such a completion is called a *local completion* and the noncontextuality condition is referred to as the *locality condition*. Theorem 2.1 translates into the fact that a behaviour in a Bell scenario has a global section if, and only if, there is a local completion recovering its statistics.

## 2.2   Probability Distributions and Physical Theories

A very natural question is, given a behaviour B, which theories could generate it? Two usual candidates are classical probability theory and quantum probability theory. Let us focus now on the set of behaviours generated by classical and quantum probabilistic models.

### 2.2.1   Classical Realisations and Noncontextuality

**Definition 2.6**  A *classical realisation* for the scenario $\Upsilon = (X, \mathscr{C}, O)$ is given by a probability space $(\Omega, \Sigma, \mu)$, where $\Omega$ is a sample space, $\Sigma$ a $\sigma$-algebra and $\mu$ a probability measure in $\Sigma$, and for each $M \in X$ a partition of $\Omega$ into $|O|$ disjoint subsets $A_j^M \in \Sigma$, $j \in O$. For each context $C = \{M_1, \ldots, M_n\}$, the probability of outcomes $s = (a_1, \ldots, a_n)$ for a joint measurement of the elements of $C$ is

$$p_C(s) = \mu \left( \bigcap_{k=1}^{n} A_{a_k}^{M_k} \right). \tag{2.11}$$

The behaviours that can be obtained in this form are called *classical* or *noncontextual behaviours*. The set of all classical behaviours will be denoted by $\mathscr{C}(\Upsilon)$.

The set $\mathscr{C}(\Upsilon)$ is a polytope with $|O^X|$ vertices. If a behaviour B is classical, we have that

$$p_U^C(s|_U) = \mu \left( \bigcap_{k | M_k \in U} A_{a_k}^{M_k} \right) \tag{2.12}$$

is independent of the context $C$, and hence B $\in \mathscr{X}(\Upsilon)$. As an immediate consequence of Theorem 2.1, we also have the following result:

**Corollary 2.1**  *A behaviour has a global section if and only if it is classical.*

In fact, once a classical realisation is given, the construction of the global section is guaranteed by the fact that the intersection of a finite number of sets in a $\sigma$-algebra also belongs to the $\sigma$-algebra. Conversely, given the global section, we can construct the classical realisation using the same argument present in the paragraph preceding Theorem 2.1. This result shows that the existence of a global section is an alternative definition for classicality, making clear its topological character.

### 2.2.2  Quantum Realisations

**Definition 2.7**  A *quantum realisation* for the scenario $\Upsilon = (X, \mathscr{C}, O)$ is given by a Hilbert space $\mathfrak{H}$, for each $M \in X$ a partition of the identity operator acting on $\mathfrak{H}$ into $|O|$ projectors $P_j^M$, $j \in O$, and a density matrix $\rho$ acting on $\mathfrak{H}$. For a given context $C = \{M_1, \ldots, M_n\} \in \mathscr{C}$, the compatibility condition demands the existence of a basis for $\mathfrak{H}$ in which all $P_j^{M_i}$ are diagonal, or, equivalently,

$$\left[ P_j^{M_i}, P_l^{M_k} \right] = 0, \forall i, j, k, l. \tag{2.13}$$

The probability of outcomes $s = (a_1, \ldots, a_n)$ for a joint measurement of $C$ is

$$p_C(s) = \mathrm{Tr}\left( \prod_{k=1}^{n} P_{a_k}^{M_k} \rho \right). \tag{2.14}$$

The behaviours that can be written in this form are called *quantum behaviours*. The set of all quantum behaviours will be denoted by $\mathscr{Q}(\Upsilon)$.

Notice that the Hilbert space is not fixed and the set $\mathscr{Q}(\Upsilon)$ contains realisations in all dimensions.

**Theorem 2.2**  *The set of quantum behaviours $\mathscr{Q}(\Upsilon)$ is a convex set.*

*Proof*  Let $\mathrm{B}^1 = \{p_C^1\}$ and $\mathrm{B}^2 = \{p_C^2\}$ be two quantum behaviours for the same scenario. We want to prove that any convex combination

$$\alpha \mathrm{B}^1 + \beta \mathrm{B}^2, \quad 0 \le \alpha, \beta \le 1, \quad \alpha + \beta = 1 \tag{2.15}$$

defined as the set of probability distributions $\{p_C = \alpha p_C^1 + \beta p_C^2\}$ is a quantum behaviour.

Let $\rho^1$ and partitions of the identity $\{P_j^{M1}\}$ be a quantum realisation for $\mathrm{B}^1$ and $\rho^2$ and partitions of the identity $\{P_j^{M2}\}$ be a quantum realisation for $\mathrm{B}^2$, that is, given $C = \{M_1, \ldots, M_n\}$ and $s = (a_1, \ldots, a_n) \in O^C$ we have

$$p_C^1(s) = \mathrm{Tr}\left( \prod_{k=1}^{n} P_{a_k}^{M_k 1} \rho^1 \right) \tag{2.16}$$

and similarly for $p_C^2$

$$p_C^2(s) = \mathrm{Tr}\left( \prod_k P_{a_k}^{M_k 2} \rho^2 \right). \tag{2.17}$$

It is important to notice here that the density matrices and projectors in the quantum realisations for $B_1$ and $B_2$ given above do not have necessarily the same dimension. Nonetheless, it is always possible to extend one of them to a Hilbert space of higher dimension, so without loss of generality we will consider that all density matrices and projectors act in the same Hilbert space $\mathfrak{H}$.

Let $\{|1\rangle, |2\rangle\}$ be an orthonormal basis for $\mathbb{C}^2$ and define the density matrix

$$\rho = \alpha\rho^1 \otimes |1\rangle\langle 1| + \beta\rho^2 \otimes |2\rangle\langle 2| \tag{2.18}$$

and the projectors

$$P_{a_k}^{M_k} = P_{a_k}^{M_k 1} \otimes |1\rangle\langle 1| + P_{a_k}^{M_k 2} \otimes |2\rangle\langle 2|, \tag{2.19}$$

acting on $\mathfrak{H} \otimes \mathbb{C}^2$. Then, we have that

$$p_C(s) := \mathrm{Tr}\left(\prod_{k=1}^{n} P_{a_k}^{M_k} \rho\right) = \alpha\left(\prod_{k=1}^{n} P_{a_k}^{M_k 1}\rho^1\right) + \beta\left(\prod_{k} P_{a_k}^{M_k 2}\rho^2\right) \tag{2.20}$$

which implies that

$$p_C = \alpha p_C^1 + \beta p_C^2. \tag{2.21}$$

Hence, any convex combination of quantum behaviours for $\Upsilon$ is also a quantum behaviour for $\Upsilon$. □

The set $\mathscr{Q}(\Upsilon)$ is convex, but it is not a polytope in general. It is also important to mention that the use of a Hilbert space of higher dimension in the previous proof cannot be avoided. In fact, if we bound the dimension of the quantum realisations, we get a set that is not convex, as shown in Ref. [PV09].

The set of classical behaviours is contained in the set of quantum behaviours. To prove that, notice that the set of behaviours obtained from a probability space with $n$ elements is equivalent to the set of behaviours obtained with diagonal projectors and density matrices in a Hilbert space of dimension $n$ with a fixed orthonormal basis. The set of elements in the sample space $\Omega$ is the set of unidimensional projectors and the measure is given by $\mu(P) = Tr(\rho P)$.

## 2.3 Noncontextuality Inequalities

We would like to find simple criteria to decide whether or not a behaviour B is noncontextual. According to Theorem 2.1, this is equivalent to test if B $\in \mathscr{C}(\Upsilon)$. We can, in principle, use the fact that $\mathscr{C}(\Upsilon)$ is a polytope to derive a finite number of inequalities that provide necessary and sufficient conditions for membership in this set.

A convex polytope may be defined as an intersection of a finite number of half-spaces. Such definition is called a *half-space representation* (H-representation or H-description). There exist infinitely many H-descriptions of a convex polytope. However, for a full-dimensional convex polytope, the minimal H-description is in fact unique and is given by the set of facet-defining halfspaces. Although the set of behaviours is not full dimensional, it is possible to apply a projection in $\mathbb{R}^d$ such that it is injective when restricted to $\mathscr{X}(\Upsilon)$ and such that $\mathscr{X}(\Upsilon)$, $\mathscr{Q}(\Upsilon)$ and $\mathscr{C}(\Upsilon)$ are full-dimensional. We will describe this projection for the scenarios where the contexts have at most two binary measurements in Sect. 2.5.

Since $\mathscr{C}(\Upsilon)$ is a polytope, there is a minimal set of inequalities giving an H-representation. Some of these inequalities are the trivial inequalities related to the definition of probability distributions (positivity and normalisation), but others are nontrivial inequalities which are, in general, not satisfied by all elements of $\mathscr{X}(\Upsilon)$. These inequalities are called *noncontextuality inequalities*.

**Definition 2.8** A *noncontextuality inequality* is a linear inequality

$$S := \sum_{s \in O^C,\, C \in \mathscr{C}} \gamma_C(s) p_C(s) \leq b, \tag{2.22}$$

where all $\gamma_C(s)$ and $b$ are real numbers, which is satisfied by all elements of the classical polytope $\mathscr{C}(\Upsilon)$ and violated by some contextual distribution. A noncontextuality inequality is called *tight* if there is some noncontextual behaviour reaching equality and it is called *facet-defining* if it defines a nontrivial facet of the classical polytope $\mathscr{C}(\Upsilon)$.

For the special case of Bell scenarios, noncontextuality inequalities are known as *Bell inequalities*.

Any H-description provides a necessary and sufficient condition for membership in $\mathscr{C}(\Upsilon)$: a behaviour B is noncontextual if and only if it satisfies all noncontextuality inequalities in this description. Although verifying if a behaviour satisfies or not each inequality is very simple, finding all inequalities that provide an H-description for a general scenario is a very difficult computational task, related to the max-cut problem, which belongs to the NP-hard class of computational complexity [BM86, DL97, AII06].

### 2.3.1  The CHSH Inequality

The CHSH scenario was introduced by Clauser, Horne, Shimony and Holt in Ref. [CHSH69]. It consists of four measurements

$$X = \{M_0, M_1, M_2, M_3\} \tag{2.23}$$

with compatibility structure given by

$$\mathscr{C} = \{\{M_0, M_1\}, \{M_1, M_2\}, \{M_2, M_3\}, \{M_3, M_0\}\}. \tag{2.24}$$

**Fig. 2.1** The compatibility hypergraph for the bipartite scenario with measurements $\{M_0, M_2\}$ (red) for the first party and measurements $\{M_1, M_3\}$ (green) for the second party

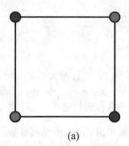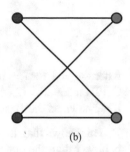

(a)          (b)

The set of possible outcomes is $O = \{\pm 1\}$. The hypergraph $\mathcal{H}$ in this case is a familiar simple graph: the square. This scenario corresponds to the Bell scenario $(2, 2, 2)$, if we regard $M_0$ and $M_2$ as measurements performed by one party and $M_1$ and $M_3$ as measurements performed by another party (Fig. 2.1).

This scenario was completely characterised in Refs. [Fin82, AQB$^+$13]. There are 8 facet-defining noncontextuality inequalities and all of them can be written in the form

$$\sum_{i=0}^{3} \gamma_i \langle M_i M_{i+1} \rangle \leq 2, \qquad (2.25)$$

where $\langle M_i M_i \rangle = p\,(M_i = M_{i+1}) - p\,(M_i \neq M_{i+1})$, $\gamma_i \in \{\pm 1\}$ and the number of $\gamma_i = -1$ is odd.

The inequality obtained with $\gamma_i - 1$ for $i = 0, 1, 2$ and $\gamma_3 = -1$ is the famous CHSH inequality, presented in the seminal paper [CHSH69]. These inequalities are violated by some quantum behaviours in dimension four or higher. The quantum maximum for inequality (2.25), also known as Tsirelson bound, is $2\sqrt{2}$. These violations can be obtained with appropriate measurements in an entangled state of two qubits[CHSH69, AQB$^+$13].

Notice that the algebraic maximum for inequality (2.25) is 4, higher than the quantum maximum. This maximum can be achieved with noncontextual behaviours, as shown in the following example.

*Example 2.3* The nondisturbing behaviour

|         | $(1, 1)$ | $(1, -1)$ | $(-1, 1)$ | $(-1, -1)$ |
|---------|----------|-----------|-----------|------------|
| $M_0 M_1$ | $\frac{1}{2}$ | $0$ | $0$ | $\frac{1}{2}$ |
| $M_1 M_2$ | $\frac{1}{2}$ | $0$ | $0$ | $\frac{1}{2}$ |
| $M_2 M_3$ | $\frac{1}{2}$ | $0$ | $0$ | $\frac{1}{2}$ |
| $M_3 M_0$ | $0$ | $\frac{1}{2}$ | $\frac{1}{2}$ | $0$ |

gives

$$\sum_{i=0}^{3} \gamma_i \langle M_i M_{i+1} \rangle = 4, \tag{2.26}$$

where $\gamma_i = 1$ for $i = 0, 1, 2$ and $\gamma_3 = -1$, reaching the algebraic maximum for the CHSH inequality (2.25). This extremal example was introduced in Ref. [PR94] and it is known as the *Popescu-Rohrlich (PR) box*.

This shows that, in general, the violation obtained with nondisturbing behaviours is higher than the quantum maximum, and hence, that the nondisturbance polytope contains properly the quantum set.

### 2.3.2   The KCBS Inequality

The KCBS scenario was introduced by Klyachko, Can, Binicioğlu and Shumovsky in Ref. [KCBS08]. It consists of five measurements

$$X = \{M_0, M_1, M_2, M_3, M_4\}, \tag{2.27}$$

with compatibility cover given by

$$\mathscr{C} = \{\{M_0, M_1\}, \{M_1, M_2\}, \{M_2, M_3\}, \{M_3, M_4\}, \{M_4, M_0\}\}. \tag{2.28}$$

The set of possible outcomes is $O = \{\pm 1\}$. The hypergraph $\mathscr{H}$ in this case is another familiar simple graph: the pentagon (Fig. 2.2).

This scenario was completely characterised in Refs. [Ara12, AQB$^+$13]. There are $2^4$ facet-defining noncontextuality inequalities and all of them can be written in the form

$$\sum_{i=0}^{4} \gamma_i \langle M_i M_{i+1} \rangle \leq 3, \tag{2.29}$$

**Fig. 2.2** The compatibility hypergraph of the KCBS scenario

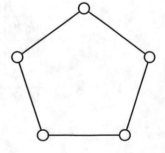

where $\langle M_i M_{i+1}\rangle = p\,(M_i = M_{i+1}) - p\,(M_i \neq M_{i+1})$, $\gamma_i \in \{\pm 1\}$ and the number of $\gamma_i = -1$ is odd.

The inequality obtained when all $\gamma_i = -1$ is the famous KCBS inequality, presented in the seminal paper [KCBS08]. It is equivalent to the inequality

$$\sum_{i=0}^{4} \langle P_i\rangle \leq 2, \tag{2.30}$$

where $M_i = 2P_i - 1$ and $\langle P_i\rangle = p\,(P_i = 1)$.

These inequalities are violated by some quantum behaviours in dimension three or higher. The maximal quantum violation for inequality (2.29) is $4\sqrt{5} - 5$, which corresponds to a maximal quantum value of $\sqrt{5}$ for inequality (2.30). These violations can be obtained with the state $|\psi\rangle = (1, 0, 0)$ and with projectors

$$P_i = \left(\cos(\theta)\ \sin(\theta)\cos\left(\tfrac{4i\pi}{5}\right)\ \sin(\theta)\sin\left(\tfrac{4i\pi}{5}\right)\right), \tag{2.31}$$

where

$$\cos^2(\theta) = \frac{\cos\left(\frac{\pi}{5}\right)}{\left(1 + \cos\left(\frac{\pi}{5}\right)\right)}. \tag{2.32}$$

An interesting property of these projectors is that $P_i$ and $P_{i+1}$ are orthogonal. This implies that the outcome $-1 - 1$ can never occur in a measurement of $M_i$ and $M_{i+1}$.

The algebraic maximum for inequality (2.29) is 5, higher than the quantum maximum. This maximum can be achieved with nondisturbing behaviours, as shown in the following example.

*Example 2.4* The nondisturbing behaviour

| | $(1, 1)$ | $(1, -1)$ | $(-1, 1)$ | $(-1, -1)$ |
|---|---|---|---|---|
| $M_0 M_1$ | $\frac{1}{2}$ | 0 | 0 | $\frac{1}{2}$ |
| $M_1 M_2$ | $\frac{1}{2}$ | 0 | 0 | $\frac{1}{2}$ |
| $M_2 M_3$ | $\frac{1}{2}$ | 0 | 0 | $\frac{1}{2}$ |
| $M_3 M_4$ | $\frac{1}{2}$ | 0 | 0 | $\frac{1}{2}$ |
| $M_4 M_0$ | 0 | $\frac{1}{2}$ | $\frac{1}{2}$ | 0 |

gives

$$\sum_{i=0}^{4} \gamma_i \langle M_i M_{i+1}\rangle = 5, \tag{2.33}$$

where $\gamma_i = 1$ for $i = 0, 1, 2, 3$ and $\gamma_4 = -1$, reaching the algebraic maximum for the KCBS inequality (2.29) for this particular choice of coefficients.

*Example 2.5* The nondisturbing behaviour

| | $(1, 1)$ | $(1, -1)$ | $(-1, 1)$ | $(-1, -1)$ |
|---|---|---|---|---|
| $M_0 M_1$ | 0 | $\frac{1}{2}$ | $\frac{1}{2}$ | 0 |
| $M_1 M_2$ | 0 | $\frac{1}{2}$ | $\frac{1}{2}$ | 0 |
| $M_2 M_3$ | 0 | $\frac{1}{2}$ | $\frac{1}{2}$ | 0 |
| $M_3 M_4$ | 0 | $\frac{1}{2}$ | $\frac{1}{2}$ | 0 |
| $M_4 M_0$ | 0 | $\frac{1}{2}$ | $\frac{1}{2}$ | 0 |

gives

$$-\sum_{i=0}^{4} \langle M_i M_{i+1} \rangle = 5, \tag{2.34}$$

reaching the algebraic maximum for the KCBS inequality (2.29) for this particular choice of coefficients.

### 2.3.3 The n-Cycle Inequalities

A simple generalisation of the CHSH and KCBS inequality is obtained when we use as compatibility hypergraph an $n$-cycle, shown in Fig. 2.3 for $n = 3, 4, 5, 6$. The $n$-cycle is a simple graph with $n$ vertices $0, 1, \ldots, n-1$ such that two vertices $i, j$ are connected iff $|i - j| = 1 \bmod n$. The corresponding scenario has $n$ measurements

$$X = \{M_0, M_1, \ldots, M_{n-1}\}, \tag{2.35}$$

with compatibility cover given by

$$\mathscr{C} = \{\{M_0, M_1\}, \{M_1, M_2\}, \ldots, \{M_{n-2}, M_{n-1}\}, \{M_{n-1}, M_0\}\}. \tag{2.36}$$

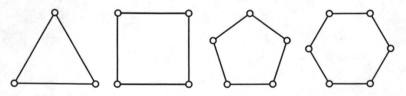

**Fig. 2.3** The compatibility hypergraph of the $n$-cycle scenario for $n = 3, 4, 5, 6$

The set of possible outcomes is also $O = \{\pm 1\}$. The complete set of noncontextuality inequalities for this scenario was found in Ref. [AQB+13].

**Theorem 2.3** *There are $2^{n-1}$ tight noncontextuality inequalities for the n-cycle scenario, and they are of the form*

$$\sum_{i=0}^{n-1} \gamma_i \langle M_i M_{i+1} \rangle \leq n - 2, \tag{2.37}$$

*where the sum is taken modulo n, $\gamma_i = \pm 1$, and the number of indices i such that $\gamma_i = -1$ is odd.*

Some quantum distributions violate this bound if $n \geq 4$. The maximum quantum violation is given by

$$\begin{cases} \frac{3n\cos\left(\frac{\pi}{n}\right) - n}{1 + \cos\left(\frac{\pi}{n}\right)} & \text{if } n \text{ is odd,} \\ n\cos\left(\frac{\pi}{n}\right) & \text{if } n \text{ is even.} \end{cases} \tag{2.38}$$

For $n$ odd, the quantum bound can be achieved already in a three-dimensional system, with the state $\begin{pmatrix} 1 & 0 & 0 \end{pmatrix}$ and measurements $M_i = 2 |v_i\rangle \langle v_i| - I$, where

$$|v_i\rangle = \left( \cos(\theta) \ \sin(\theta)\cos\left(\frac{i\pi(n-1)}{n}\right) \ \sin(\theta)\sin\left(\frac{i\pi(n-1)}{n}\right) \right) \tag{2.39}$$

and

$$\cos^2(\theta) = \frac{\cos\left(\frac{\pi}{n}\right)}{\left(1 + \cos\left(\frac{\pi}{n}\right)\right)}. \tag{2.40}$$

For $n$ even, the quantum bound can be achieved in a four-dimensional system, with the state

$$|\Psi\rangle = \left( 0 \ \tfrac{1}{\sqrt{2}} \ -\tfrac{1}{\sqrt{2}} \ 0 \right) \tag{2.41}$$

and measurements $M_i = X_i \otimes I$ for odd $i$ and $M_i = I \otimes X_i$ for even $i$, where

$$X_i = \cos\left(\frac{i\pi}{n}\right)\sigma_x + \sin\left(\frac{i\pi}{n}\right)\sigma_z. \tag{2.42}$$

These bounds were calculated in Ref. [AQB+13] with the help of the tools we will introduce in the next section.

The n-cycle scenarios are, in some sense, the simplest ones exhibiting quantum noncontextual behaviours. They were also the first ones to be fully characterised. Besides that, the interest in these scenarios comes also from the fact that if the compatibility graph does not contain a cycle, all behaviours are noncontextual.

**Definition 2.9** Given a graph G, let $S \subset V$ be any subset of vertices of G. Then, the *induced subgraph* G $[S]$ is the graph whose vertex-set is $S$ and whose edge-set consists of all of the edges in E (G) that connect any two vertices in $S$.

**Theorem 2.4** *There is a quantum contextual behaviour in the scenario* $\Upsilon$ *if, and only if, the compatibility graph of the scenario* $\mathscr{G}$ *has an n-cycle as induced subgraph with* $n > 3$.

Equivalently, we may say that there is quantum violation of some noncontextuality inequality for the scenario if, and only if, $\mathscr{G}$ has an $n$-cycle as induced subgraph with $n > 3$. In this sense, the $n$-cycle scenarios are the simplest ones where it is possible to find quantum violations of noncontextuality inequalities. For a proof of this result, see Ref. [BM10].

## 2.4   The Exclusivity Graph

Given a scenario $(X, \mathscr{C}, O)$, it is possible to define yet another graph related to it that allows the calculation of several bounds for the associated noncontextuality inequalities. We introduce some definitions first. In what follows,

$$a_1, \ldots, a_n | M_1, \ldots, M_n \tag{2.43}$$

will denote the measurement event where the compatible measurements in context $C = \{M_1, \ldots, M_n\}$ were performed and outcomes $s = (a_1, \ldots, a_n) \in O^C$ were observed. We will also use the alternative notation

$$p(a_1, \ldots, a_n | M_1, \ldots, M_n) := p_C(s). \tag{2.44}$$

**Definition 2.10** We say that two measurement events

$$a_1, \ldots, a_n | M_1, \ldots, M_n \text{ and } a'_1, \ldots, a'_n | M'_1, \ldots, M'_n \tag{2.45}$$

are *exclusive* if for some $i$ and $j$, $M_i = M'_j$ and $a_i \neq a'_j$, that is, the same measurement is performed in both contexts and different outcomes are observed.

**Definition 2.11** The *exclusivity graph* $\mathfrak{G}$ of the scenario $(X, \mathscr{C}, O)$ is the simple graph whose vertices are labelled by all possible measurement events

$$a_1, \ldots, a_n | M_1, \ldots, M_n, \quad M_i \in X, \ C = \{M_1, \ldots, M_n\} \in \mathscr{C}, \ (a_1, \ldots, a_n) \in O^C \tag{2.46}$$

and two vertices are connected by an edge if, and only if, the corresponding events are exclusive.

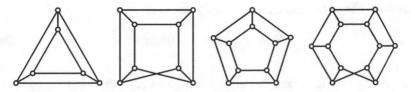

**Fig. 2.4** Exclusivity graphs for the $n$-cycle inequalities for $n = 3, 4, 5, 6$

Generally, not all possible measurement events appear in a given inequality. The ones that do appear define an induced subgraph of $\mathfrak{G}$ that can give many insights about the inequality itself.

**Definition 2.12** The *exclusivity graph* G of a noncontextuality inequality is the induced subgraph of $\mathfrak{G}$ defined by the vertices that correspond to measurement events appearing in the inequality.

*Example 2.6 (The Exclusivity Graphs of the n-Cycle Inequalities)* Since

$$\langle M_i M_j \rangle = 2 \left( p(11|M_i M_j) + p(-1-1|M_i M_j) \right) - 1 \tag{2.47}$$

and

$$-\langle M_i M_j \rangle = 2 \left( p(1-1|M_i M_j) + p(-11|M_i M_j) \right) - 1, \tag{2.48}$$

there are $2n$ events in each noncontextuality inequality (2.37) for the $n$-cycle scenario. If $n$ is odd, the corresponding exclusivity graph is $Y_n$, the *prism graph* of order $n$, and if $n$ is even, the exclusivity graph is $M_{2n}$, the *Möbius ladder* of order $2n$. The first four of these graphs are depicted in Fig. 2.4.

Let us now consider the case where all coefficients $\gamma_C(s)$ in inequality (2.22) are equal to one. Many important inequalities can be written in this form, including the $n$-cycle inequalities. In this case, we can use the exclusivity graph G and some graph functions to derive the maximal bounds for the quantity $S$ in inequality (2.22) in different probabilistic theories. First, a few definitions from graph theory.

**Definition 2.13** An *independent set* or *stable set* in a graph G is a set of vertices of G, no two of which are adjacent. A *maximum independent set* is an independent set of largest possible size for G.

**Definition 2.14** The *independence number* $\alpha$ (G) of a graph G is the cardinality of a maximum independent set of G.

**Definition 2.15** Let $\{1, \ldots, n\}$ be the set of vertices of a graph G. An *orthonormal representation* for G in a finite-dimensional vector space with inner product V is a set of unit vectors $\{|u_1\rangle, \ldots, |u_n\rangle\}$ such that $|u_i\rangle$ and $|u_j\rangle$ are orthogonal whenever $i$ and $j$ are not connected in G.

**Definition 2.16** The *Lovász number* of a graph $G$ is

$$\vartheta(G) = \sup \sum_i \in V(G) |\langle u_i \mid \psi \rangle|^2 \tag{2.49}$$

where the supremum is taken over all vector spaces $V$, all orthogonal representations $\{|u_1\rangle, \ldots, |u_n\rangle\}$ for $\overline{G}$ and all unit vectors $|\psi\rangle \in V$.

An orthonormal representation achieving the supremum, called an *optimal orthonormal representation*, always exists, as shown in [Lov79].

Both $\alpha(G)$ and $\vartheta(G)$ play fundamental roles in the graph approach to contextuality. For a more detailed discussion about these graph functions, see [Lov79, Lov95, Knu94, Ros67, Bol98].

**Theorem 2.5 (Cabello, Severini and Winter, 2010)** *The classical bound of a noncontextuality inequality is the independence number $\alpha(G)$ of the exclusivity graph G of the inequality.*

*Proof* Since noncontextuality inequalities are linear, the maximum classical bound is achieved in an extremal point of the noncontextual polytope. For such a point, the probability of each event is either zero or one and the value of the sum $S$ for this behaviour is equal to the number of events with probability one. Since the sum of the probabilities of two exclusive events cannot be higher than one, two connected vertices cannot have probability equal to one at the same time. Hence, the set of vertices whose probabilities are one is an independent set, and hence can not have more than $\alpha(G)$ elements.

To prove that equality holds, it suffices to take any maximum independent set and use a classical behaviour that assigns probability one to each event in this set.    □

**Theorem 2.6 (Cabello, Severini and Winter, 2010)** *The quantum bound of a noncontextuality inequality is upper bounded by the Lovász number $\vartheta(G)$ of the exclusivity graph G of the inequality.*

*Proof* The maximal quantum value for $S$ is obtained with a pure state $\rho = |\psi\rangle \langle\psi|$. Let $\{e_i\}$, $e_i = a_0^i, \ldots, a_{n_i}^i |M_1^i \ldots, M_{n_i}^i$, be the set of events appearing in the inequality and $P_i = \prod_k P_{a_k^i}^{M_k^i}$ be the projector corresponding to $e_i$, where $P_{a_k^i}^{M_k^i}$ is the projector associated to outcome $a_k^i$ for measurement $M_k^i$. Define

$$|v_i\rangle = \frac{P_i |\psi\rangle}{|P_i |\psi\rangle|}. \tag{2.50}$$

Then, we have

$$S = \sum_i p(a_0^i, \ldots, a_{n_i}^i | M_1^i, \ldots, M_{n_i}^i) = \sum_{i \in V(G)} |\langle \psi \mid v_i \rangle|^2. \tag{2.51}$$

If $e_i$ and $e_j$ are exclusive events, the corresponding projectors $P_i$ and $P_j$ are orthogonal, and hence $|v_i\rangle$ and $|v_j\rangle$ are also orthogonal. The set of vectors $|v_i\rangle$ and the state $|\psi\rangle$ provide an orthogonal representation for $\overline{G}$, which implies

$$\sum_{i \in V(G)} |\langle \psi \mid v_i \rangle|^2 \leq \vartheta(G). \tag{2.52}$$

□

*Example 2.7 (Quantum Bound for the n-Cycle Inequalities)* The observation that

$$\vartheta(Y_n) = \frac{3n \cos\left(\frac{\pi}{n}\right) - n}{1 + \cos\left(\frac{\pi}{n}\right)}, \quad \vartheta(M_{2n}) = n \cos\left(\frac{\pi}{n}\right) \tag{2.53}$$

and Theorem 2.6 were used by the authors in Ref. [AQB$^+$13] to find the quantum maximum violation of the $n$-cycle inequalities, as shown in Sect. 2.3.3.

Although in the previous example the bound was tight, this is not true in general. This can happen when the scenario imposes extra constraints that make the Lovász optimal representations for the graph unattainable for quantum systems.

*Example 2.8* In Ref. [SBBC13], the authors present three inequalities for which $\vartheta(G)$ is larger than the quantum maximum. Consider the scenario where the system is composed by two spatially separated parties. In the first subsystem, there are two measurements available, denoted by $A_0$ and $A_1$, and in the second subsystem we also have two measurements available, denoted by $B_0$ and $B_1$. All measurements have two possible outputs, 0 and 1. In this case, the compatibility of the measurements in different systems is guaranteed by spatial separation. The compatibility hypergraph is a square, with edges linking measurements in different parties, as shown in Fig. 2.1.

This scenario admits two noncontextuality inequalities with quantum bound larger than the classical bound for which the exclusivity graph is a pentagon:

$$p(00|00) + p(11|01) + p(10|11) + p(00|10) + p(11|00) \leq 2, \tag{2.54}$$

$$p(00|00) + p(11|01) + p(10|11) + p(00|10) + p(\_1|\_0) \leq 2. \tag{2.55}$$

In the inequalities above, $ab|xy$ denotes the measurement event where the first party applies measurement $A_x$ and gets outcome $a$ and the second party applies measurement $B_y$ and gets outcome $b$; $\_1|\_0$ corresponds to the event where the second party applies measurement $B_0$ and gets outcome 1, irrespectively of the first party's action.

The quantum bound for the first inequality is approximately 2.178, while for the second it is approximately 2.207. The events appearing in these inequalities and their exclusivity structures are shown Fig. 2.5.

Consider the scenario whose compatibility hypergraph is shown in Fig. 2.6. This scenario admits one noncontextuality inequality with quantum bound larger than the classical bound for which the exclusivity graph is a pentagon:

$$p(00|00) + p(11|01) + p(10|11) + p(00|10) + p(11|20) \leq 2. \tag{2.56}$$

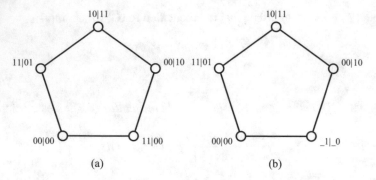

**Fig. 2.5** The labelling of the exclusivity graph for the Bell inequalities (2.55) with pentagonal exclusivity structure

**Fig. 2.6** The compatibility hypergraph for the bipartite scenario with measurements $\{A_0, A_1, A_2\}$ (red) for the first party and measurements $\{B_0, B_1\}$ (green) for the second party

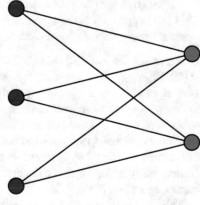

**Fig. 2.7** The labelling of the exclusivity graph for the Bell inequalities (2.56) with pentagonal exclusivity structure

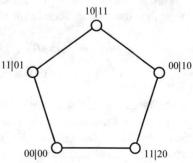

The quantum bound for this inequality is approximately 2.207. The events appearing in these inequalities and their exclusivity structure are shown in Fig. 2.7.

For each of these inequalities, the quantum bound is strictly smaller than the Lovász number of the pentagon $\vartheta\,(C_5) = \sqrt{5} \approx 2.236$. This proves that, in general, $\vartheta\,(G)$ gives only a loose upper bound for the maximum quantum value of the inequality.

### 2.4.1 Vertex-Weighted Exclusivity Graph

Theorems 2.5 and 2.6 relate noncontextual and quantum bounds for

$$S := \sum_{s \in O^C, C \in \mathscr{C}} \gamma_C(s) p_C(s) \tag{2.57}$$

with graph invariants when the coefficients $\gamma_C(s)$ are all equal to one. If this is not the case, we can get analogous results using the weighted versions of $\alpha$ and $\vartheta$.

First, we notice that, without loss of generality, we can assume that $\gamma_C(s) > 0$ for all $C \in \mathscr{C}$ and $s \in O^C$. In fact, if $\gamma_C(s) < 0$ for some $C \in \mathscr{C}$ and $s \in O^C$, we use normalisation of probability distributions to rewrite $S$ replacing $p_C(s)$ by

$$p_C(s) = 1 - \sum_{s' \neq s} p_C(s'). \tag{2.58}$$

With this assumption, we define $\gamma$ as the vector in $\mathbb{R}^{|V(G)|}$ with positive coordinates given by $\gamma_C(s)$.

**Definition 2.17** The *vertex-weighted exclusivity graph* $(G, \gamma)$ of a noncontextuality inequality

$$S := \sum_{s \in O^C, C \in \mathscr{C}} \gamma_C(s) p_C(s) \leq b \tag{2.59}$$

is the vertex-weighted graph defined by the exclusivity graph G of the inequality and the vertex weights given by the coefficients $\gamma_C(s)$ of the inequality.

**Definition 2.18** The *weighted independence number* of a vertex-weighted graph $(G, \gamma)$ is

$$\alpha(G, \gamma) = \max_W \sum_{i \in W} \gamma_i \tag{2.60}$$

where the maximum is taken over all independent sets $W \subset V(G)$.

**Definition 2.19** The *vertex-weighted Lovász number* of a vertex-weighted graph $(G, \gamma)$ is

$$\vartheta(G) = \max \sum_{i \in V(G)} \gamma_i |\langle u_i | \psi \rangle|^2 \tag{2.61}$$

where the maximum is taken over all vector spaces $V$, all orthogonal representations $\{|u_1\rangle, \ldots, |u_n\rangle\}$ for $\overline{G}$ and all unit vectors $|\psi\rangle \in V$.

Notice that $\alpha(G, \gamma)$ and $\vartheta(G, \gamma)$ coincide with $\alpha(G)$ and $\vartheta(G)$, respectively, when $\gamma_i = 1$ for all $i \in V(G)$.

**Theorem 2.7 (Cabello, Severini and Winter, 2010)** *The classical bound of a noncontextuality inequality is the independence number $\alpha\,(G, \gamma)$ of the vertex-weighted exclusivity graph $(G, \gamma)$ of the inequality. The quantum bound is upper bounded by the vertex-weighted Lovász number $\vartheta\,(G, \gamma)$ of $(G, \gamma)$.*

The proofs are analogous to those of Theorems 2.5 and 2.6. The reader can also find more details in Refs. [CSW10, CSW14], including a prescription of how to reach $\vartheta\,(G, \gamma)$ within quantum theory.

## 2.5 The Geometry of the Case $\mathcal{H} = \mathcal{G}$

For the scenarios $\Upsilon$ in which every context consists of at most two measurements, the compatibility hypergraph $\mathcal{H}$ is equal to the compatibility graph $\mathcal{G}$. If each measurement has two outcomes, labelled 0 and 1, both nondisturbing and noncontextual sets can be equivalently described in different ways that lead to familiar polytopes from graph theory [AII06, AT17].

### 2.5.1 Description of the Nondisturbing, Quantum and Noncontextual Behaviours

In this type of scenario, the nondisturbing set $\mathscr{X}\,(\Upsilon)$ is a subset of $\mathbb{R}^{4|E(G)|}$. Given a context $\{M_i, M_j\} \in \mathscr{C}$, we denote by $p_{ij}(ab)$ the probability of obtaining outcome $a$ for measurement $i$ and outcome $b$ for measurement $j$. We denote by $p_i(a) = \sum_b p_{ij}(ab)$ the marginal probability for measurement $M_i$ and by $p_i(b) = \sum_a p_{ij}(ab)$ the marginal probability for measurement $M_j$.

The conditions imposed on the nondisturbing behaviours B allow us to determine all its entries knowing only $p_{ij}(11)$ and $p_i(1)$. In fact, we can define

$$\phi : \mathbb{R}^{4|E(G)|} \longrightarrow \mathbb{R}^{|V(G)|+|E(G)|}$$

$$B \longmapsto q = \left(q_i, q_{kj}\right)_{i \in V(G); (k,j) \in E(G)} \tag{2.62}$$

where $q$ is such that $q_i = p_i(1)$ and $q_{ij} = p_{ij}(11)$. To recover B from $q$, just notice that

$$p_{ij}(10) = q_i - q_{ij}$$

$$p_{ij}(01) = q_j - q_{ij}$$

$$p_{ij}(00) = 1 - q_i - q_j + q_{ij}. \tag{2.63}$$

Notice that this map is injective when restricted to $\mathscr{X}\,(\Upsilon)$. It happens that the image of all nondisturbing behaviours for this scenario under the action of transformation $\phi$ is equal to a well-known convex polytope from graph theory, the *correlation polytope* of G.

**Definition 2.20**  Given $S \subset \mathrm{V}(G)$, we define the *correlation vector* $v(S) \in \mathbb{R}^{|V(G)|+|E(G)|}$ as

$$v(S)_i = \begin{cases} 1 & \text{if } i \in S; \\ 0 & \text{otherwise,} \end{cases} \quad \forall i \in \mathrm{V}(G), \tag{2.64}$$

$$v(S)_{ij} = \begin{cases} 1 & \text{if } i, j \in S; \\ 0 & \text{otherwise,} \end{cases} \quad \forall \{i, j\} \in \mathrm{E}(G). \tag{2.65}$$

The *correlation polytope* COR (G) is the convex hull of all correlation vectors.

Notice that the correlation vectors correspond to the image of the extremal behaviours in $\mathscr{C}(\Upsilon)$ under the action of $\phi$, $S$ being the set of vertices in $\mathrm{V}(G)$ with certain outcome equal to 1 [AII06, AT17]. This proves the following result:

**Theorem 2.8**  $\phi(\mathscr{C}(\Upsilon)) = \mathrm{COR}(G)$.

The image of the nondisturbance polytope is also a well-known polytope from graph theory.

**Definition 2.21**  The *rooted correlation semimetric polytope* RCMET (G) of a graph G is the set of vectors $q = (q_i, q_{jk}) \in \mathbb{R}^{|V(G)|+|E(G)|}$ such that

$$q_{ij} \geq 0, \tag{2.66}$$

$$q_i - q_{ij} \geq 0, \tag{2.67}$$

$$1 - q_i - q_j + q_{ij} \geq 0. \tag{2.68}$$

**Theorem 2.9**  $\phi(\mathscr{X}(G)) = \mathrm{RCMET}(G)$.

*Proof*  As a consequence of Eqs. (2.63), (2.66) is equivalent to positivity of $p_{ij}$ (11); Eq. (2.67) is equivalent to positivity of $p_{ij}$ (10) and, if we exchange the roles of $i$ and $j$, to positivity of $p_{ij}$ (01). Finally, Eq. (2.68) is equivalent to positivity of $p_{ij}$ (00). Hence, $q \in \mathrm{RCMET}(G)$ if and only if the vector $p$ recovered from $q$ using Eq. (2.63) belongs to $\mathscr{X}(G)$, which implies $\phi(\mathscr{X}(G)) = \mathrm{RCMET}(G)$. ☐

### 2.5.2   The Cut Polytope

**Definition 2.22**  Given a graph G and $c \in \{0, 1\}^{|V(G)|}$, the *cut vector* of G defined by $c$ is the vector $x(c) \in \mathbb{R}^{|E(G)|}$ such that

$$x(c)_{ij} = c_i \oplus c_j, \tag{2.69}$$

where $\oplus$ denotes sum modulo 2. The *cut polytope* of G, denoted by $\mathrm{CUT}^{01}(G)$, is the convex hull of all cut vectors of G.

There exists a relation between the polytopes CUT and COR.

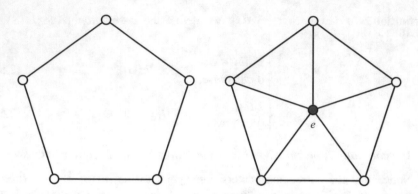

**Fig. 2.8** The pentagon $C_5$ and its suspension graph $\nabla C_5$

**Definition 2.23** The *suspension graph* $\nabla G$ of G is the graph with vertex-set $V(G) \sqcup \{e\}$ and edge-set $E(G) \cup \{(e, i), i \in V(G)\}$.

Intuitively, $\nabla G$ is the graph obtained from G by adding an extra vertex and connecting it to all vertices of G. The suspension graph of the pentagon $C_5$ is shown in Fig. 2.8.

**Theorem 2.10** $CUT^{\pm 1}(\nabla G) = \psi(COR(G))$, *in which*

$$\psi : \mathbb{R}^{|V(G)|+|E(G)|} \longrightarrow \mathbb{R}^{|V(G)|+|E(G)|} \tag{2.70}$$

$$q \longmapsto x \tag{2.71}$$

*and the coordinates of x are given by*

$$\begin{aligned} x_{ij} &= 1 - q_i - q_j + 2q_{ij}, \quad \{i, j\} \in E(G) \\ x_{ei} &= q_i, \qquad\qquad\qquad i \in V(G) \end{aligned} \tag{2.72}$$

For a proof of this result, see Refs. [DL97, AII06, AT17].

The entries $x_{ei}$ of the vector $x$ are the expectation values of the measurements $i$:

$$x_{ei} = \langle i \rangle = p_i(1) \tag{2.73}$$

and the entries $x_{ij}$ of $x$ is the probability that the outcomes of measurements $i$ and $j$ are equal; that is, when $i$ and $j$ are compatible measurements, we have:

$$x_{ij} = p(i = j) = p_{ij}(00) + p_{ij}(11). \tag{2.74}$$

In the particular case where $i$ and $j$ are perfectly correlated, we have $x_{ij} = 1$ and if they are perfectly anti-correlated we have $x_{ij} = 0$.

We can also define the cut polytope using the output values $-1$ and $1$ instead of $0$ and $1$.

**Definition 2.24** Given a graph G and $c \in \{-1, 1\}^{|V(G)|}$, the $\pm1$-*cut vector* of G defined by $c$ is the vector $y(c) \in \mathbb{R}^{|E(G)|}$ such that

$$y(c)_{ij} = c_i c_j. \tag{2.75}$$

The $\pm1$-*cut polytope* of G, $\text{CUT}^{\pm1}$ (G), is the convex hull of all $\pm1$-cut vectors of G.

The two definitions $\text{CUT}^{\pm1}$ and $\text{CUT}^{01}$ are related by a bijective linear map

$$\alpha : \text{CUT}^{01} (G) \longrightarrow \text{CUT}^{\pm1} (G) \tag{2.76}$$

$$x \longmapsto y \tag{2.77}$$

$$y_{ij} = 1 - 2x_{ij}. \tag{2.78}$$

If we denote the outcomes of each measurement by $\pm1$, the entries $y_{ei}$ of the vector $y$ are the expectation values of the measurements $i$:

$$y_{ei} = \langle i \rangle = p_i(1) - p_i(-1) \tag{2.79}$$

while the entries $y_{ij}$ of $y$ are the expectation values of the products $ij$, when $i$ and $j$ are compatible measurements:

$$y_{ij} = \langle ij \rangle = p(i = j) - p(i \neq j)$$
$$= p_{ij}(1, 1) + p_{ij}(-1, -1) - p_{ij}(-1, 1) - p_{ij}(1, -1). \tag{2.80}$$

These polytopes are very hard to characterise for general scenarios, because the number of extremal points grows enormously with the number of vertices in G. Hence, it may be useful to look for connections between them and other simpler polytopes, even if these connections are only valid for a restricted class of graphs. Following this idea, for some graphs it is possible to relate $\text{CUT}^{01}$ (G) with the so-called *semimetric polytope* of $\mathcal{G}$, denoted by MET (G).

**Definition 2.25** Let V be a set and

$$U(V) = \{\{i, j\} \mid i, j \in V\} \tag{2.81}$$

be the set of unordered pairs of elements of V. The *semimetric polytope*[3] MET on V is the set of vectors in $\mathbb{R}^{|U(V)|}$ satisfying

$$x_{ij} \geq 0 \;\; \forall \{i, j\} \in U(V), \tag{2.82}$$

$$x_{ij} - x_{ik} - x_{kj} \leq 0 \;\; \forall i, j, k \in V. \tag{2.83}$$

The *semimetric polytope* of a graph G, MET (G), is the projection of the semimetric polytope on V (G) over the entries corresponding to the unordered pairs $\{i, j\} \in E(G)$.

---

[3]The semimetric polytope, as the name suggests, is related to the notion of semimetrics defined on the set V. For a detailed discussion about MET, see Ref. [DL97].

Inequalities (2.83) are called *triangular inequalities*.

**Definition 2.26** An *edge contraction* is an operation which removes an edge from a graph while simultaneously merging the two vertices it used to connect. A graph $G'$ is a *minor* of another graph G if it can be obtained from G by contracting edges, deleting edges and deleting isolated vertices.

**Theorem 2.11** $CUT^{01}$ (G) = MET (G) *if, and only if, G has no $K_5$ minor.*

This result is extremely useful since MET (G) is easily characterised by the following result:

**Theorem 2.12** *Given $F \subset E (G)$ and $x \in \mathbb{R}^{|E(G)|}$, let*

$$x(F) = \sum_{(i,j)\in F} x_{ij}. \tag{2.84}$$

*The following are true for* MET (G):

1. MET (G) $= \{x \in \mathbb{R}^{|E(G)|} |\ x_{ij} \leq 1,\ x(F) - x(C \setminus F) \leq |F| - 1,\ C$ *cycle of* $G, F \subset C, |F|$ *odd*$\}$;
2. *The inequality $x(F) - x(C \setminus F) \leq |F| - 1$ defines a facet of* MET (G) *if, and only if, C is an induced cycle of* G;
3. *The inequality $x_{ij} \leq 1$ defines a facet of* MET (G) *if, and only if, the edge $(i, j)$ does not belong to a triangle of* G.

Theorems 2.11 and 2.12 can be used to find all facets of $CUT^{01}$ (G) if G has no $K_5$-minor. In this case, the facets are defined by the so-called *n-cycle inequalities*:

$$x(F) - x(C \setminus F) \leq |F| - 1,\ C \text{ cycle of } G, F \subset C, |F| \text{ odd.} \tag{2.85}$$

We can use these inequalities and the map $\alpha$ to find the facet-defining inequalities of $CUT^{\pm 1}$ (G), if G has no $K_5$-minor, which are given by

$$y(F) - y(C \setminus F) \leq |C| - 2,\ C \text{ cycle of } G, F \subset C, |F| \text{ odd.} \tag{2.86}$$

This is the same set of inequalities found for the special case $G = C_n$ in Ref. [AQB$^+$13].

The map $\psi$ that relates the correlation polytope and the cut polytope also relates the polytope RCMET (G) to other known polytope, the *rooted semimetric polytope* of G.

**Definition 2.27** Let $e$ be an extra vertex of the suspension graph $\nabla G$ with respect to G. The subset of triangle inequalities (2.83) corresponding to all triples $(e, i, j)$ for all $(i, j) \in E (G)$ defines the rooted semimetric polytope RMET (G).

**Theorem 2.13** *The image of* RCMET (G) *under* $\psi$ *is the* rooted semimetric polytope *of* $\nabla G$, RMET ($\nabla G$).

The proofs of Theorems 2.11, 2.12, 2.13 and many other properties of these polytopes, which are certainly useful in the study of contextuality and non-locality for this particular scenario, can be found in Refs. [DL97, AT17].

### 2.5.3 Correlation Functions

To describe completely the sets $\mathcal{X}(\Upsilon)$, $\mathcal{Q}(\Upsilon)$ and $\mathcal{C}(\Upsilon)$ using the composition of the maps $\phi$ and $\psi$, we have to use vectors in $\mathbb{R}^{|V(G)|+|E(G)|}$. In some situations, it might be useful to consider a projection of these vectors in $\mathbb{R}^{|E(G)|}$, obtained by eliminating the coordinates relative to the edges $(e, i)$:

$$\Pi : \mathbb{R}^{|V(G)|+|E(G)|} \longrightarrow \mathbb{R}^{|E(G)|} \tag{2.87}$$

$$x = (x_i, x_{jk})_{i \in |V(G)|; (j,k) \in |E(G)|} \longmapsto (x_{jk})_{(j,k) \in |E(G)|}. \tag{2.88}$$

The vectors in $\Pi$ (RMET (G)) are called *correlation functions*.[4]

**Theorem 2.14** *Given a graph G, the following are true:*

1. $\Pi \,(\mathrm{RMET}\,(G)) = [0, 1]^{|E(G)|}$;
2. $\Pi \,(\mathrm{CUT}^{01}\,(\nabla G)) = \mathrm{CUT}^{01}\,(G)$.

See Ref. [DL97] for a proof. Notice that this projection is not injective when restricted to $\mathcal{X}(\Upsilon)$. The knowledge of the correlation function is not enough to fully recover the behaviour, since we are loosing the information on the marginals when we apply the projection $\Pi$. Nonetheless, these vectors may be useful for two reasons: first, they provide a simpler description of the behaviours, which give some information in scenarios where $\nabla G$ is too complicated to deal with; second, although correlation functions do not give full information about the behaviour, they can be enough to decide whether the corresponding behaviours are *contextual or not*. This happens, for example, for the $n$-cycle scenarios [AQB$^+$13].

### 2.5.4 The Eliptope and the Set of Quantum Behaviours

For Bell scenarios with two parties, one with $n$ measurements at her disposal and the other with $m$ measurements at his disposal, the corresponding compatibility graph is the complete bipartite graph $K_{m,n}$. In this particular type of scenario, with the outcomes of each measurement labelled $\pm 1$, the set $\Pi(\mathcal{Q})$ is related to the *eliptope* of $K_{m,n}$.

**Theorem 2.15 (Tsirelson)** *The following are true:*

1. $y = (\langle ij \rangle) \in \Pi(\mathcal{Q})$;

---

[4]Correlation functions are also known in the literature as *correlation vectors* or simply *correlators*. We avoid this nomenclature to distinguish this definition from Definition 2.20.

2. *There are vectors* $|u_i\rangle$, $|v_j\rangle \in \mathbb{R}^d$, $1 \le i \le m$, $1 \le j \le n$, $d \le m + n$, *such that*

$$y_{ij} = \langle u_i \mid v_j \rangle.$$  (2.89)

**Definition 2.28** The *eliptope* $\mathscr{E}$ (G) of a graph G is the set of vectors $x \in \mathbb{R}^{|E(G)|}$ such that for each $i \in V$ (G) there is a unit vector $|u_i\rangle \in \mathbb{R}^{|V(G)|}$ such that

$$y_{ij} = \langle u_i \mid u_j \rangle.$$  (2.90)

With this definition, Tsirelson's theorem states that the set of quantum correlation vectors in a bipartite Bell scenario is the eliptope of $K_{m,n}$. For a proof of this result, see Ref. [AII06].

The natural question is whether Tsirelson's theorem is also valid for general contextuality scenarios, that is, we want to know if given any graph G, the equality $\Pi(\mathscr{Q}(G)) = \mathscr{E}(G)$ holds. The inclusion $\Pi(\mathscr{Q}(\Upsilon)) \subset \mathscr{E}(\Upsilon)$ is always true, as shown in Ref. [AT17].

**Theorem 2.16** $\Pi(\mathscr{Q}(\Upsilon)) \subset \mathscr{E}$ (G).

For some graphs, the inclusion $\mathscr{E}$ (G) $\subset \Pi$ ($\mathscr{Q}$(G)) does not hold. This is the case for the $n$-cycle $C_n$ for any odd $n$. This is shown by the fact that the violation of the $n$-cycle inequalities for some points in the eliptope can be larger than the maximum violation obtained with quantum behaviours.

**Theorem 2.17** *There is a point* $y \in \mathscr{E}$ ($C_n$) *for which*

$$\sum_{i=0}^{n-2} y_{ii+1} - y_{0n-1} = n \cos\left(\frac{\pi}{n}\right).$$  (2.91)

This point is explained in detail in Ref. [AT17]. Its existence proves that, in general, $\Pi(\mathscr{Q}(\Upsilon)) \ne \mathscr{E}$ (G). For any odd $n$, Proposition 2.17 shows that there is an element for which $\sum_{i=0}^{n-2} y_{ii+1} - y_{0n-1} = n \cos\left(\frac{\pi}{n}\right)$, while the quantum maximum for this same quantity is

$$\frac{3n \cos\left(\frac{\pi}{n}\right) - n}{1 + \cos\left(\frac{\pi}{n}\right)} < n \cos\frac{\pi}{n}.$$  (2.92)

Another family for which $\mathscr{E}$ (G) is different from the quantum set is the family of complete graphs $K_n$. In this case, all measurements are compatible and hence the quantum set is equal to the classical set, a polytope. On the other hand, $\mathscr{E}$ (G) is a polytope if, and only if, G is a forest [DL97], and in this case $CUT^{\pm 1}$ (G) = $\mathscr{E}$ (G) = $\Pi$ (RMET (G)) = $[-1, 1]^{|E(G)|}$.

For the $n$-cycles with $n$ even, $\mathscr{E}$ ($C_n$) = $\Pi(\mathscr{Q}(\Upsilon))$. This is a consequence of the fact that in this case $C_n$ is a subgraph of the complete bipartite graph $K_{n/2,n/2}$ and the eliptope of $C_n$ is a projection of the eliptope of $K_{n/2,n/2}$.

## 2.6   Sheaf Theory and Contextuality

It is possible to provide a beautiful mathematical formulation of contextuality using categories and sheaf theory, as pioneered by Abramsky and Brandenburger [AB11]. This approach provides a direct and unified characterisation of both contextuality and non-locality, along with different new tools, insights and results. We provide a brief introduction to the sheaf-theoretical aspects of contextuality in this section and we refer to Refs. [AB11, Man13] for more detailed definitions and discussions.

We start with a contextuality scenario $\Upsilon = (X, \mathscr{C}, O)$. When a set of compatible measurements $C \in \mathscr{C}$ is performed, a set of outcomes in $O^C$ will be observed. Measurement events in $O^C$ and functions

$$s : C \longrightarrow O \tag{2.93}$$

are in bijective correspondence. Such a function is called a *section* over $C$.

Define the function $\varepsilon$ that takes each context $C$ to $O^C$, the set of sections over $C$. We can also define a natural action by restriction according to Sect. 2.1: if $C \subseteq C'$, let

$$r_C^{C'} : \varepsilon(C') \longrightarrow \varepsilon(C)$$

$$s \longmapsto s|_C. \tag{2.94}$$

This restriction is such that

$$r_C^C = id_{\varepsilon(C)} \tag{2.95}$$

and if $C \subseteq C' \subseteq C''$,

$$r_C^{C'} \circ r_{C'}^{C''} = r_C^{C''}. \tag{2.96}$$

Let **Set** be the category whose objects are sets and arrows are functions between sets. Let $\mathscr{P}(X)$ be the category whose objects are the subsets of $X$ and there is a unique arrow from $C$ to $C'$ if and only if $C \subseteq C'$. Let $\mathscr{P}(X)^{op}$ be the category whose objects are the subsets of $X$ and there is a unique arrow from[5] $C'$ to $C$ if and only if $C \subseteq C'$. Then, we can see the function $\varepsilon$ defined above as a functor[6]

$$\varepsilon : \mathscr{P}(X)^{op} \longrightarrow \textbf{Set} \tag{2.97}$$

that takes each $C \subseteq X$ to $\varepsilon(C) = O^C$ and the unique arrow $C' \to C$ to the restriction $r_C^{C'}$, when $C \subseteq C'$. Equations (2.95) and (2.96) prove that $\varepsilon$ is in fact a functor and hence $\varepsilon$ is a *presheaf*.

---

[5]The *opposite category* or *dual category* $\mathscr{C}^{op}$ of a given category $\mathscr{C}$ is formed by reversing the arrows, that is, interchanging the source and target of each arrow [MM92, Mac98].

[6]More precisely, we have a different functor for each choice of $O$, but for simplicity we will not make this dependence explicit in the notation.

**Definition 2.29** Given a category $D$, a functor $F : D^{op} \to \mathbf{Set}$ is called a *presheaf*.

The functor $\varepsilon$ has another distinguished property. Let $C \subseteq X$ and $\{C_i\}_{i \in I}$ be a family of subsets of $C$ such that $\bigcup_i C_i = C$ and $\{s_i \in \varepsilon(C_i)\}_{i \in I}$ be a family of sections that agree in all intersections, that is

$$s_i|_{C_i \cap C_j} = s_j|_{C_i \cap C_j} \tag{2.98}$$

for every $i, j \in I$. Then, there is a unique global section $s \in \varepsilon(C)$ such that $s|_{C_i} = s_i$. In fact, given $M \in C$ there is at least one $i \in I$ such that $M \in C_i$. Let $m = s_i(M)$. Since all sections $s_i$ agree on the overlaps, $m$ does not depend on the index $i$ chosen. We define then $s(M) = m$.

This distinguished property is called the *sheaf condition* and $\varepsilon$ is called the *sheaf of events*.

**Definition 2.30** Let $F : \mathscr{P}(X)^{op} \to \mathbf{Set}$ be a presheaf and $f_C^{C'} : F(C') \to F(C)$ be the arrow in $\mathbf{Set}$ associated to the unique arrow $C' \to C$ if $C \subseteq C'$. If $s \in F(C')$, let $s|_C = f_C^{C'}(s)$. We say that $F$ is a *sheaf* if it satisfies the following two conditions:

1. Locality: If $(C_i \subseteq X)_{i \in I}$ is a covering of $C \subseteq X$, and if $s, t \in F(C)$ are such that $s|_{C_i} = t|_{C_i}$ for each set $C_i$, then $s = t$;
2. Gluing: If $(C_i)_{i \in I}$ is a covering of $C$, and if for each $i$ there is a section $s_i$ over $C_i$ such that for each pair $i, j$,

$$s_i|_{C_i \cap C_j} = s_j|_{C_i \cap C_j}, \tag{2.99}$$

there is a section $s \in F(C)$ such that $s|_{C_i} = s_i$ for each $i$.

Sections correspond to definite outcomes, but most of the times it is not possible to predict with certainty the outcome of every measurement. When probabilistic theories enter the game, we must use *probability distributions* over the set of sections $O^C$. To make definitions more general, we will consider distributions taking values over a commutative semiring $R$.

**Definition 2.31** An *R-distribution* on a set U is a function $d : \mathrm{U} \to R$ such that

$$\sum_{u \in \mathrm{U}} d(u) = 1. \tag{2.100}$$

When we are interested in probability distributions, $R$ is the semiring of positive real numbers. Nonetheless, it is quite instructive to keep $R$ general, even when we are working with probabilities in a compatibility scenario.[7] We denote by $\mathscr{D}_R(C)$ the set of $R$-distributions on $C$.

---

[7]For our purposes, $R$ will be the semiring of non-negative real numbers. For a nice use of "negative probabilities", see Ref. [Man13].

Let $f : U \to V$ be a surjective function between two sets U and V. We define

$$\mathscr{D}_R(f) : \mathscr{D}_R(U) \longrightarrow \mathscr{D}_R(V) \tag{2.101}$$

that takes each distribution $d : U \to R$ to the distribution $\mathscr{D}_R(f)(d) = d' : V \longrightarrow R$ defined by

$$d'(v) = \sum_{u;\, f(u)=v} d(u). \tag{2.102}$$

This definition is functorial since $\mathscr{D}_R(id) = id$ and $\mathscr{D}_R(g \circ f) = \mathscr{D}_R(g) \circ \mathscr{D}_R(f)$. With this definition, we can construct the functor

$$\mathscr{D}_R : \textbf{Set} \to \textbf{Set} \tag{2.103}$$

that takes each set U to the set of $R$-distributions on $U$ and each function $f : U \to V$ to the function $\mathscr{D}_R(f) : \mathscr{D}_R(U) \longrightarrow \mathscr{D}_R(V)$.

The composition of the functor $\mathscr{D}_R$ with the sheaf $\varepsilon$ defines the presheaf

$$\mathscr{D}_R \circ \varepsilon : \mathscr{P}(X)^{op} \longrightarrow \textbf{Set}, \tag{2.104}$$

which assigns to each subset $C \subseteq X$ the set of $R$-distributions on the sections over $C$. If $C \subseteq C'$, the unique arrow $C' \to C$ is taken by this presheaf to the restriction map $r_C^{C'}$ acting on the set of $R$-distribution on $O^{C'}$: if $d \in \mathscr{D}_R(\varepsilon(C'))$, then

$$r_C^{C'}(d) = d|_C \tag{2.105}$$

where $d|_C(s) = \sum_{s';\, s'|_C = s} d(s')$.

The restriction $d|_C$ is the *marginal* distribution of $d$, which assigns to each section $s$ in the smaller set $C$ the sum of the weights of all sections $s'$ in the larger set $C'$ that restrict to $s$.

With the language of category theory introduced above, a behaviour[8] for the scenario $(X, \mathscr{C}, O)$ is a family of $R$-distributions $p_C \in \mathscr{D}_R(\varepsilon(C))$, $C \in \mathscr{C}$. Once more, we consider only nondisturbing behaviours, that is, we demand that

$$p_C|_{C \cap C'} = p_{C'}|_{C \cap C'} \tag{2.106}$$

whenever $C \cap C' \neq \emptyset$.

We have already observed that the presheaf $\varepsilon$ is indeed a sheaf. It is natural to ask if the same holds for the presheaf $\mathscr{D}_R \circ \varepsilon$. The nondisturbance condition corresponds precisely to the locality condition in Definition 2.30. Then, any nondisturbing behaviour fulfils the preconditions for gluing. The gluing condition, on the other hand, is not always satisfied, since it requires the existence of a global distribution

---

[8] In Ref. [AB11], behaviours are referred to as *empirical models*.

$d \in \mathscr{D}_R \circ \varepsilon(X)$ such that $d|_C = p_C$ to each context $C$. Theorem 2.1 implies that such a distribution $d$ exists if and only if there is a completion reproducing the statistics of the behaviour.

**Theorem 2.18** *A behaviour $B = \{p_C\}$ satisfies the sheaf condition if and only if there is a completion reproducing its statistics.*

There are some compatibility scenarios for which the presheaf $\mathscr{D}_R \circ \varepsilon$ does satisfy the gluing condition, which implies that it is impossible to find examples of contextuality on these scenarios. These compatibility structures are characterised by Vorob'ev theorem [Vor62]. The interested reader can find a discussion of this result in the language of contextuality in Ref. [Bar14].

### 2.6.1 Bundle Diagrams

The characterisation of contextuality in terms of the failure of the sheaf condition can be displayed in a visually appealing *bundle diagram*. In this diagram, the *base space* represents the measurements in $X$ and there is an edge between two measurements when they are compatible. Above each vertex, there is a *fibre* with the possible outcomes of each measurement, that is, the elements of $O$. Given a behaviour B, we draw an edge between values in adjacent fibres if the corresponding joint outcome is possible, that is, if the probability assigned to this joint outcome by B is larger than zero. Notice that the diagrams display the *possibilistic* structure of the behaviour: it only distinguishes zero probabilities from positive probabilities.

A global section $s \in O^X$ corresponds to a closed path traversing all the fibres exactly once. Then, if there is a global section $s$ consistent with the behaviour, the lifting of a closed path in X defined by $s$ in the fibres is also closed.

*Example 2.9* The nondisturbing behaviour $B$ defined by the table

|          | $(1, 1)$      | $(1, -1)$ | $(-1, 1)$ | $(-1, -1)$    |
|----------|---------------|-----------|-----------|---------------|
| $M_0 M_1$ | $\frac{1}{2}$ | 0         | 0         | $\frac{1}{2}$ |
| $M_1 M_2$ | $\frac{1}{2}$ | 0         | 0         | $\frac{1}{2}$ |
| $M_2 M_3$ | $\frac{1}{2}$ | 0         | 0         | $\frac{1}{2}$ |
| $M_3 M_4$ | $\frac{1}{2}$ | 0         | 0         | $\frac{1}{2}$ |
| $M_4 M_0$ | $\frac{1}{2}$ | 0         | 0         | $\frac{1}{2}$ |

is a noncontextual behaviour, with global probability distribution satisfying

$$p_X(-1, -1, -1, -1, -1) = \frac{1}{2} \tag{2.107}$$

$$p_X(1, 1, 1, 1, 1) = \frac{1}{2}. \tag{2.108}$$

**Fig. 2.9** Bundle diagram of
the noncontextual behaviour
$B$ of Example 2.9

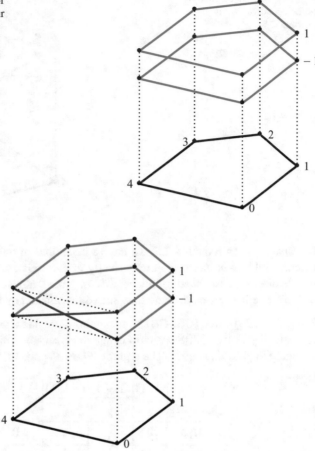

**Fig. 2.10** The bundle diagram of the behaviour shown in Example 2.4. The base, in black, consists
of the compatibility graph of the scenario. Each fibre contains the possible outcomes of each
measurement, in this case $-1$ and $1$. We connect two values in adjacent fibres when that joint
outcome has probability larger than zero. For $i = 0, \ldots, 3$, $M_i$ and $M_{i+i}$ have always equal
outcomes, corresponding to the green edges between neighbour fibres. Measurements $M_4$ and $M_0$
have always opposite outcomes, corresponding to the red edges between the fibres

The associated bundle diagram is shown in Fig. 2.9. Notice that in this case every
edge between adjacent fibres belongs to a closed path transversing the fibres only
once (one of the green pentagons).

*Example 2.10* In Example 2.4, we have a behaviour in the KCBS scenario where
measurements $M_i$ and $M_{i+1}$ are perfectly correlated for $i = 0, \ldots 3$, while $M_4$ and
$M_0$ are perfectly anti-correlated. Figure 2.10 shows the bundle diagram associated
to this behaviour. Note that there is no closed curve in the bundle using only the
edges allowed by the behaviour.

**Fig. 2.11** Bundle diagram of
the behaviour shown in
Example 2.5

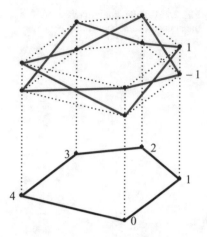

*Example 2.11* In Example 2.5, we have a behaviour in the KCBS scenario where
all compatible measurements are perfectly anti-correlated. Figure 2.11 shows the
bundle diagram associated to this behaviour. Note that also in this case there is no
closed curve in the bundle using only the edges allowed by the behaviour.

*Example 2.12* In Ref. [CBTCB13], the authors provide a simple proof of quantum
contextuality in the KCBS scenario using an argument similar to the one proposed
by Hardy in [Har92, Har93]. The argument corresponds to the table

|          | $(1, 1)$     | $(1, -1)$    | $(-1, 1)$    | $(-1, -1)$ |
|----------|--------------|--------------|--------------|------------|
| $M_0 M_1$ | $\frac{2}{9}$ | $\frac{2}{3}$ | $\frac{1}{9}$ | $0$        |
| $M_1 M_2$ | $0$          | $\frac{1}{3}$ | $\frac{2}{3}$ | $0$        |
| $M_2 M_3$ | $\frac{1}{3}$ | $\frac{1}{3}$ | $\frac{1}{3}$ | $0$        |
| $M_3 M_4$ | $0$          | $\frac{2}{3}$ | $\frac{1}{3}$ | $0$        |
| $M_4 M_0$ | $\frac{2}{9}$ | $\frac{1}{9}$ | $\frac{2}{3}$ | $0$        |

associated to a noncontextual quantum behaviour. The corresponding bundle diagram is shown in Fig. 2.12.

Notice that in Example 2.12, some of the liftings of the base pentagon are closed,
while others are not. In Examples 2.10 and 2.11, there are no closed liftings, an
extreme situation known as *strong contextuality* [AB11].

With this, we have a characterisation of the phenomena of contextuality in terms
of *obstructions to the existence of global sections in a presheaf*, that can be computed using a linear algebraic approach. With this formalism, the authors in [AB11]
distinguish a proper hierarchy of strengths of contextuality, and show three leading
examples that occupy successively higher levels of this hierarchy. They also prove a
general correspondence between the existence of noncontextual completions using

**Fig. 2.12** Bundle diagram of noncontextual behaviour $B$ of Example 2.12

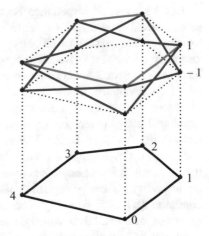

negative probabilities and nondisturbing behaviours. Another approach to identify these obstructions using cohomology was proposed in Refs. [AMB11, ABK$^+$15]. There, the authors define some invariants of a topological space related to a behaviour, which are non-zero only if the behaviour is contextual. These results open the door to the use of the methods of sheaf theory to the study of contextuality, which not only provide an elegant mathematical description of the problem but also lead to new tools and insights.

## 2.7 Final Remarks

The connection between contextuality and graph theory has led to many developments in the field. In this chapter, we have focused on the compatibility-hypergraph approach, the usual approach to quantum contextuality, where an experimentalist is given a set of possible measurements to perform in a physical system, and the compatibility structure of this set is encoded in the *compatibility hypergraph* of the scenario.

The connection between graphs and contextuality provides practical tools for the characterisation of the sets of noncontextual, quantum and nondisturbing behaviours. The noncontextual set is a polytope and the quantum set is in general larger, as proven by the fact that some quantum behaviours do not satisfy all noncontextuality inequalities that bound the noncontextual polytope. This proves the impossibility of noncontextual completions of quantum theory, i.e. Bell-Kochen-Specker theorem.

The mathematical formalism of this scenario can be translated into a sheaf-theoretical language, which provides an elegant characterisation of the phenomena of contextuality in terms of *obstructions to the existence of global sections in a presheaf*, which opens the door to the use of the methods of topology to the study of contextuality.

Besides its clear importance for foundations of physics, the graph approach to contextuality is useful for more operational applications of contextuality. For example, when the compatibility hypergraph is equal to the compatibility graph, the geometry of the sets of nondisturbing, quantum and noncontextual behaviours we have seen in Sect. 2.5 can be explored in the definition of contextuality quantifiers based on geometric distances [AT17, BAC17]. This has important applications in resource theories of contextuality [HGJ$^+$15, GHH$^+$14, ACTA17, ABM17], developed after the recognition of contextuality as a potential resource for several tasks [VVE12, Rau13, UZZ$^+$, HWVE14, DAGBR15, JBVR17, SS17, SHP17]. The compatibility-hypergraph approach and its geometrical aspects are also crucial for the characterisation of *extended contextuality*, a generalisation of the usual notion of contextuality that can be applied to behaviours that do not satisfy the nondisturbance condition [KDL15, ADO17]. The advantage of this extended notion is that it allows experimental data, which usually do not satisfy rigorously the nondisturbing condition, to be better confronted with noncontextual extended bounds.

Another important tool in this approach is the *exclusivity graph* of a noncontextuality inequality. The noncontextual bound is given by the weighted independence number and the quantum bound is upper bounded by the weighted Lovász number of this graph. As we show in Sect. 2.4, the upper bound is not always achieved when the compatibility structure is imposed. In the next chapter, we discuss the exclusivity-graph approach to contextuality, where only exclusivity relations among measurement events are taken into account. This offers a wider spectrum of interesting phenomena. We will see that when there are no other restriction on the measurement events, the Lovász number can always be achieved using quantum resources. We will also see that another graph invariant can be defined in order to take such extra restrictions into account.

# Chapter 3
# Contextuality: The Exclusivity-Graph Approach

The mathematical content of the original proof of the Bell-Kochen-Specker theorem states that there are sets of one-dimensional projectors for which it is not possible to assign definite values 0 or 1 noncontextually in such a way that if a set of mutually orthogonal projectors add to identity, then the value 1 must be assigned to one, and only one, of them (for more details, see Sect. A.3). The set of one-dimensional projectors in a proof of the Bell-Kochen-Specker theorem can be represented using a graph, known as the *Kochen-Specker diagram*. The vertices of the graph correspond to the projectors and two of them are joined by an edge whenever the corresponding projectors are compatible, which, in this case, also implies that they are orthogonal.

The usual physical interpretation of this result connects each projector to a measurement in a quantum system with possible outcomes 0 and 1. The non-contextuality assumption translates into the observation that the value assigned to each measurement is independent of other compatible measurements performed jointly. With this association, the theorem implies the impossibility of noncontextual assignment of definite values to all measurements in a quantum system consistently with the quantum statistics, proving the impossibility of noncontextual completions of quantum theory.

We can look at this result from a different perspective. Instead of associating projectors to measurements, we can associate projectors to *measurement events*. With this interpretation, each vertex in a Kochen-Specker diagram corresponds to an outcome of a measurement and two vertices are connected by an edge if they are exclusive, that is, if they are associated to two different outcomes of one and the same measurement. Several examples of Kochen-Specker diagrams will be presented in Appendix A.

A completion for quantum theory provides definite values to all measurements, and hence, given a vertex $P$ in the Kochen-Specker diagram, we know if the measurement event it corresponds to occurs or not. If the outcome associated to the measurement by the completion is the one that corresponds to $P$, we associate the value 1 to $P$. Otherwise, we associate the value 0 to $P$.

B. Amaral, M. Terra Cunha, *On Graph Approaches to Contextuality and their Role in Quantum Theory*, SpringerBriefs in Mathematics, https://doi.org/10.1007/978-3-319-93827-1_3

If we have a set of projectors $\{P_1, \ldots, P_n\}$ summing up to identity, we know that there is a measurement for which exclusive outcomes are associated to these projectors. Hence, since one, and only one, outcome must occur, one, and only one, of these projectors is associated to the value 1. Hence, we have

$$\sum_i v(P_i) = 1, \tag{3.1}$$

where $v(P_i)$ is the value assigned by the completion to projector $P_i$.

In this new perspective, the noncontextuality condition entails that the value associated to a projector $P$ by the completion is independent of the other projectors used to define the measurement.

**Assumption 3.1 (Measurement Noncontextuality)** *Whenever P corresponds to an outcome of different measurements* $M_1, M_2, \ldots, M_n$, *a noncontextual completion assigns the outcome corresponding to P to some* $M_i$ *if, and only if, it does for all other* $M_j$.

We can also see the KCBS inequality discussed in Sect. 2.3.2 in this new perspective. The compatibility graph G of this scenario is a pentagon and the maximum quantum violation is obtained with projectors $P_i$ such that $P_i$ and $P_j$ are orthogonal if $\{i, j\} \in E(G)$. This observation leads to two different interpretations of the graph G in quantum realisations of this particular case. In the first one, each vertex $i$ of G can be seen as the observable associated to the projector $P_i$. In the second one, each vertex $i$ of G is seen as the measurement event associated to $P_i$, and for each edge $\{i, j\}$ there must be a measurement $M$ which includes outcomes associated to $P_i$ and $P_j$. In the first interpretation, G is the compatibility graph of the set of measurements, discussed in Chap. 2. In the second interpretation, G is the *exclusivity graph* of the set of measurement events. This approach and the consequences of the assumption of measurement contextuality are the subject of the present chapter.

## 3.1 The Exclusivity Graph

In the *exclusivity-graph approach*, we start with a graph G with vertex-set V(G) and edge-set E(G). For each $i \in V(G)$, we associate a measurement event $e_i$ in a probabilistic model and for each $\{i, j\} \in E(G)$ a measurement $M$ such that $e_i$ and $e_j$ are associated to different outcomes of $M$, that is, the measurement event $e_i$ corresponds to *measuring M and observing outcome a*, whereas the measurement event $e_j$ corresponds to *measuring M and observing outcome b*, with $a \neq b$.

**Definition 3.1** Two measurement events $e_i$ and $e_j$ are *exclusive* if they are associated to different outcomes of the same measurement.

With this definition, the measurement events $e_i$ and $e_j$ associated with vertices $i$ and $j$ of G are exclusive whenever $i$ and $j$ are connected in G.

In any probabilistic theory, each state of the system provides a probability $p_i = p(e_i)$ for the occurrence of each measurement event $e_i$. We collect all these probabilities $p_i$ in a vector $p \in \mathbb{R}^{|V(G)|}$.

**Definition 3.2**  A *behaviour* for the exclusivity graph G is a map

$$p : V(G) \longrightarrow [0, 1] \tag{3.2}$$

that assigns to each $i \in V(G)$ a probability $p_i$, such that

$$p_i + p_j \leq 1, \ \forall \{i, j\} \in E(G). \tag{3.3}$$

The set of behaviours depends on the exclusivity constraints imposed by the graph G and also on the physical theory used to describe the system under study. Below, we give a beautiful description of this set for classical probability theory, quantum theory and generalised probabilistic theories with certain properties, explained in detail in Sects. 3.1.3 and 3.1.4.

### 3.1.1   Classical Noncontextual Behaviours

**Definition 3.3**  A behaviour $p$ for the exclusivity graph G is *classical* or *noncontextual* if there exist a probability space $(\Omega, \Sigma, \mu)$, where $\Omega$ is a sample space, $\Sigma$ a $\sigma$−algebra in $\Omega$ and $\mu$ a probability measure in $\Sigma$, and for each $i \in V(G)$ a set $A_i \in \Sigma$ such that $A_i \cap A_j = \varnothing$ if $\{i, j\} \in E(G)$ and

$$p_i = \mu(A_i). \tag{3.4}$$

The set of classical behaviours is called the *classical set* or *noncontextual set* and is denoted by $\mathscr{E}_C(G)$.

The behaviours outside the classical set are called *contextual*. The set $\mathscr{E}_C(G)$ is a polytope and, incidentally, it is a well-known convex polytope in the computer science literature, the *stable set polytope*, denoted by STAB(G) [Ros67, GLS86, GLS93, Knu94].

**Definition 3.4**  A *characteristic vector* for a subset $S \subset V(G)$ is a vector $p \in \mathbb{R}^{|V(G)|}$ such that for each $i \in V(G)$ the corresponding coordinate $p_i$ is given by

$$p_i = \begin{cases} 1, & \text{if } i \in S, \\ 0, & \text{if } i \notin S. \end{cases} \tag{3.5}$$

**Definition 3.5**  The *stable set polytope* STAB(G) of a graph G is the convex hull of all characteristic vectors of stable sets of G.

**Theorem 3.1**  *The set $\mathscr{E}_C(G)$ is equal to the stable set* STAB(G).

*Proof* The extremal points of $\mathscr{E}_C$ (G) are obtained with extremal measures $\mu$, for which we have $\mu(A_i) \in \{0, 1\}$ for each $A_i$. If $i$ and $j$ are connected, the associated measurable sets $A_i$ and $A_j$ are disjoint, and hence cannot both have measure one. This implies that the set

$$S = \{i \in V \mid \mu(A_i) = 1\} \tag{3.6}$$

is a stable set and $p$ is the characteristic vector of $S$, proving that $\mathscr{E}_C$ (G) $\subset$ STAB (G).

On the other hand, let $S$ be a stable set in $G$ and $p$ be its characteristic vector. Construct a classical behaviour in the following way: take the sample space $\Omega = \{1, \dots, n\}$ where $n = |V(G) \setminus S| + 1$, the $\sigma$-algebra $\Sigma = \mathscr{P}(\Omega)$ and the measure $\mu$ such that $\mu(1) = 1$ and $\mu(k) = 0$ for $k > 1$. Associate the vertices in $i \in S$ with the measurable set $\{1\}$ and the vertices $i \in V(G) \setminus S$ with each of the $|V(G) \setminus S|$ measurable sets $\{k\}$ with $k > 1$. The classical behaviour obtained with this construction is exactly the characteristic vector $p$, which proves that STAB (G) $\subset \mathscr{E}_C$ (G).                                                  $\square$

**Theorem 3.2** *Let $p$ be a behaviour for G obtained with some probabilistic model. Then, $p$ can be also be obtained with a completion of this probabilistic model satisfying measurement noncontextuality if, and only if, $p$ is classical.*

This result is another consequence of the proof of Theorem 3.1. Indeed, the extremal behaviours in any completion satisfying measurement noncontextuality must be exactly the extremal points of STAB (G).

The stable set polytope is deeply connected to the *stable set problem*, which belongs to the NP-hard complexity class. This polytope contains facets that have, in general, a highly complex structure. Therefore, one would not expect to find a "simple" set of facet-defining linear inequalities characterising the classical polytope for general graphs. By "simple" we mean a set of inequalities for which the validity of all of them can be verified in polynomial time. If such a set existed, this would prove the equivalence of the complexity classes NP and *co*-NP, which most researchers in the field strongly distrust [GLS93, PS93].

### 3.1.2  Quantum Behaviours

**Definition 3.6** A behaviour $p$ for the exclusivity graph G is *quantum* if there exist a density matrix $\rho$ acting on a Hilbert space $\mathfrak{H}$ and for each $i \in V(G)$ a projector $P_i$ acting on $\mathfrak{H}$ such that $P_i$ and $P_j$ are orthogonal if $\{i, j\} \in E(G)$ and

$$p_i = \mathrm{Tr}(P_i \rho). \tag{3.7}$$

The set of quantum behaviours is called the *quantum set* and will be denoted by $\mathscr{E}_Q$ (G).

If we fix a basis for $\mathfrak{H}$ and consider only the matrices that are diagonal in this basis, we recover the set of classical behaviours. Hence,

$$\mathscr{E}_C(G) \subset \mathscr{E}_Q(G). \tag{3.8}$$

Unlike $\mathscr{E}_C(G)$, the quantum set is not a polytope in general. This set is a well-known convex body in the computer science literature, the *theta body*, denoted by TH (G) [GLS93, Knu94, Ros67].

We recall that an orthonormal representation for a graph G is an assignment of unit vectors $|v_i\rangle \in \mathbb{R}^d$ to each vertex $i \in V(G)$ such that $\langle v_i \mid v_j \rangle = 0$ whenever $\{i, j\} \notin E(G)$. The number $d$ is called the *dimension* of the orthogonal representation. In other words, whenever we have $|v_i\rangle$ and $|v_j\rangle$ not perpendicular in the orthonormal representation, we must have $\{i, j\} \in E(G)$.

**Definition 3.7** The *cost of a vector* $|v_i\rangle$ in an orthonormal representation is defined as

$$c_i = |\langle \psi \mid v_i \rangle|^2 \tag{3.9}$$

where $|\psi\rangle = (1, 0, \ldots, 0)$ is the vector in $\mathbb{R}^d$ with the first coordinate equal to 1 and all others equal to 0.

**Definition 3.8** The *theta body* of a graph G is defined as

$$\mathrm{TH}(G) = \left\{ p \in \mathbb{R}^{|V(G)|} \;\middle|\; p_i = c_i \right\} \tag{3.10}$$

where $c_i$ is the cost of vector $|v_i\rangle$ in an orthonormal representation $\{|v_i\rangle\}$ of $\overline{G}$, the complementary graph of G.

Notice that if $\{|v_i\rangle\}$ is an orthonormal representation for $\overline{G}$, $\langle v_i \mid v_j \rangle = 0$ whenever $\{i, j\} \in E(G)$.

**Theorem 3.3** *The set* $\mathscr{E}_Q(G)$ *is equal to the theta body* TH (G).

*Proof* Given an orthonormal representation $\{|v_i\rangle\}$ for $\overline{G}$, let $P_i = |v_i\rangle \langle v_i|$. The projectors $P_i$ and $P_j$ are orthogonal if $i$ and $j$ are connected in G and hence $P_i$ and $\rho = |\psi\rangle \langle \psi|$ provide a quantum behaviour for G such that

$$p_i = \mathrm{Tr}(P_i \rho) = |\langle \psi \mid v_i \rangle|^2 = c_i. \tag{3.11}$$

This proves that TH (G) $\subset \mathscr{E}_Q(G)$.

To prove that $\mathscr{E}_Q(G) \subset \mathrm{TH}(G)$, it suffices to prove that the extremal points of the first are contained in the latter. Hence, we can assume that the state $\rho$ that defines the quantum behaviour is pure, that is, $\rho = |\phi\rangle \langle \phi|$ for some vector $|\phi\rangle$. Changing basis if necessary, we can assume that $|\phi\rangle = |\psi\rangle$. Let $P_i$ be the projector associated to vertex $i$ in the quantum behaviour. Define

$$|v_i\rangle = \frac{P_i \, |\psi\rangle}{\| P_i \, |\psi\rangle \|}. \tag{3.12}$$

If $i$ and $j$ are connected by an edge in G, the corresponding projectors are orthogonal and the vectors $|v_i\rangle$ and $|v_j\rangle$ are also orthogonal. Hence, the set of vectors $|v_i\rangle$ provide an orthonormal representation for $\overline{G}$ such that

$$c_i = |\langle \psi \mid v_i\rangle|^2 = \frac{|\langle \psi| P_i |\psi\rangle|^2}{|\langle \psi| P_i |\psi\rangle|} = \mathrm{Tr}\,(P_i |\psi\rangle \langle \psi|) = \mathrm{Tr}\,(P_i \rho) = p_i. \qquad (3.13)$$

Therefore, any extremal quantum behaviour gives a vector $p \in \mathrm{TH}\,(G)$, which implies $\mathscr{E}_Q\,(G) \subset \mathrm{TH}\,(G)$ and concludes the proof.                                     □

Unlike the classical set, the characterisation of $\mathrm{TH}\,(G)$ can be done efficiently, using semidefinite programming[1] [Knu94, GLS93].

### 3.1.3   The Exclusivity Principle

One of the most fascinating challenges in the current research in foundations of quantum theory is to provide simple physical principles that single out quantum theory in the landscape of general probabilistic theories. With this purpose in mind, we will also consider behaviours obtained when we use generalised probability theories to calculate the probabilities $p_i$, but with the restriction that they satisfy the following principle:

**Principle 3.1 (The Exclusivity Principle)** *Given a set $\{e_i\}$ of pairwise exclusive measurement events, the corresponding probabilities $p_i$ satisfy*

$$\sum_i p_i \leq 1. \qquad (3.14)$$

From the graph-theoretical point of view, this restriction is equivalent to impose the condition that whenever the set of vertices $\{i\}$ is a clique[2] in G, the sum of the corresponding probabilities $p_i$ cannot exceed one. From now on, we refer to the Exclusivity principle simply as the *E-principle*.

Specker pointed out that, in quantum theory, pairwise joint measurability of a set $\mathscr{M}$ of observables implies joint measurability of $\mathscr{M}$, while in other theories this implication does not need to hold [Spe60]. This property is known as the *Specker principle*. Later, Specker conjectured that this is *the fundamental theorem* of quantum theory [Spe]. The E-principle is a consequence of the Specker principle, as shown in Ref. [NBD+13].

---

[1]It is quite intriguing that the classical set be something algorithmically complex, while the quantum set is simple. Could it be the case that nature prefer quantum theory due to its (algorithmic) simplicity?

[2]A clique in G is a complete induced subgraph of G.

The E-principle can be used to explain why (some) distributions outside the quantum set are forbidden. Many promising results were found so far, as we discuss in Chap. 4.

### 3.1.4 E-Principle Behaviours

**Definition 3.9** A behaviour $p$ for the exclusivity graph G is an *E-principle behaviour* if the corresponding probabilities $p_i$ satisfy the E-principle. The set of E-principle behaviours is called the *E-principle set* and will be denoted by $\mathscr{E}_E$ (G).

The set $\mathscr{E}_E$ (G) is also a polytope, known in the computer science literature as the *clique-constrained stable set polytope*, or *fractional stable set polytope*, and denoted by QSTAB (G) [GLS93, Knu94, Ros67].

**Definition 3.10** The *clique-constrained stable set polytope* of a graph G is defined as

$$\text{QSTAB (G)} = \left\{ p \in \mathbb{R}^{|V(G)|} \,\middle|\, p_i \geq 0, \sum_{i \in Q} p_i \leq 1 \text{ for every clique } Q \subset V\,(G) \right\}.$$
(3.15)

**Theorem 3.4** *The E-principle set* $\mathscr{E}_E$ (G) *is equal to* QSTAB (G).

*Proof* A subset of vertices $Q \subset V$ (G) is a clique in the exclusivity graph G if, and only if, the corresponding set of measurement events is pairwise exclusive. Hence, the conditions $p_i \geq 0$ and $\sum_{i \in Q} p_i \leq 1$ are exactly equivalent to the constraints that each $p_i$ represents a probability and that the E-principle is satisfied. □

It is a known fact from the computer science literature that TH (G) $\subset$ QSTAB (G), which is equivalent to $\mathscr{E}_Q$ (G) $\subset$ $\mathscr{E}_E$ (G). This was also proven in Refs. [CSW14, FSA+13].

**Theorem 3.5** *Quantum behaviours satisfy the E-principle.*

*Proof* In quantum theory, exclusive events are associated to orthogonal projectors. Hence, if $\{e_i\}$ is a set of mutually exclusive events, a quantum realisation will provide a set $\{P_i\}$ of mutually orthogonal projectors. As a consequence, we have

$$\sum_i P_i \leq I$$
(3.16)

and hence

$$\sum_i p_i = \sum_i Tr\,(P_i \rho) \leq Tr\,(\rho) \leq 1.$$
(3.17)

□

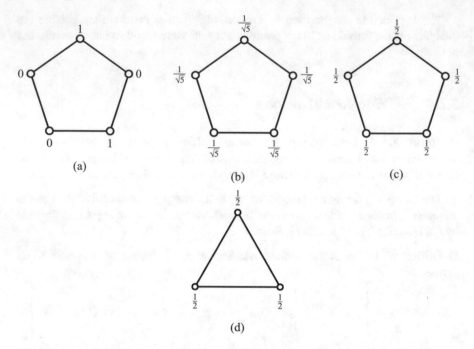

**Fig. 3.1** (**a**) A classical behaviour. (**b**) A quantum behaviour. (**c**) An E-principle behaviour. (**d**) A behaviour not satisfying the E-principle

Figure 3.1 shows examples of the different kinds of behaviours defined above.

## 3.2   Noncontextuality Inequalities

Since the classical set $\mathscr{E}_C$ (G) is a polytope, it admits an H-description: a finite set of linear inequalities which provides a necessary and sufficient condition for membership in this set. Although a complete characterisation of these inequalities is not feasible for general graphs, we can at least find simple necessary conditions for membership in this set, related to graph invariants of G.

**Definition 3.11** A *noncontextuality inequality* is a linear inequality

$$\sum_{i \in V(G)} \gamma_i p_i \leq \beta, \tag{3.18}$$

where all $\gamma_i$ and $\beta$ are real numbers, which is satisfied by all elements of the classical polytope $\mathscr{E}_C$ (G) and is violated some contextual behaviour. A noncontextuality inequality is *tight* if it is saturated for some classical behaviour and *facet-defining* if it defines a nontrivial facet of the classical polytope $\mathscr{E}_C(G)$.

To obtain necessary and sufficient conditions for membership in $\mathscr{E}_C(G)$, we have to find all facet-defining noncontextuality inequalities for G. As we mentioned before, this is a difficult problem for general graphs. Nonetheless, it may be useful to concentrate in one particular inequality and find out what information it can give.

Given a graph G, consider, for example, the sum of probabilities

$$S = \sum_{i \in V(G)} p_i. \tag{3.19}$$

We can use this sum to provide necessary conditions for membership in $\mathscr{E}_C(G)$, $\mathscr{E}_Q(G)$ and $\mathscr{E}_E(G)$. To derive these conditions, we need to identify what are the maximum values of $\beta$ in inequality (3.18) for classical, quantum and E-principle behaviours, which will be denoted, respectively, by $\beta_C$, $\beta_Q$ and $\beta_E$. Naturally, by Theorem 3.5 and the fact that $\mathscr{E}_C(G) \subset \mathscr{E}_Q(G)$, we have

$$\beta_C \le \beta_Q \le \beta_E. \tag{3.20}$$

The inequality

$$\sum_{i \in V(G)} p_i \le \beta_C \tag{3.21}$$

is a *noncontextuality inequality* as long as $\beta_C < \beta_E$ and

$$\sum_{i \in V(G)} p_i \le \beta_Q \tag{3.22}$$

is a necessary condition for membership in $\mathscr{E}_Q(G)$.

Also in the exclusivity-graph approach, the graph functions $\alpha(G)$ and $\vartheta(G)$ can be used to calculate $\beta_C$ and $\beta_Q$. The bound $\beta_E$ can be calculated with the help of another graph function, known as the *fractional-packing number* of G.

**Definition 3.12** The *fractional-packing number* $\alpha^*(G)$ of a graph G is defined as

$$\alpha^*(G) = \sup \left\{ \sum_i p_i \,\middle|\, 0 \le p_i \le 1 \text{ and } \sum_{i \in Q} p_i \le 1, Q \text{ any clique of } G \right\}. \tag{3.23}$$

**Theorem 3.6 (Cabello, Severini and Winter, 2010)** *Given a graph G, we have*

$$\beta_C = \alpha(G), \quad \beta_Q = \vartheta(G), \quad \beta_E = \alpha^*(G) \tag{3.24}$$

*where $\alpha(G)$ is the independence number of G, $\vartheta(G)$ is the Lovász number of G and $\alpha^*(G)$ is the fractional-packing number of G.*

This result follows directly from the observation that $\mathscr{E}_C(G) = \text{STAB}(G)$, $\mathscr{E}_Q(G) = \text{TH}(G)$ and $\mathscr{E}_E(G) = \text{QSTAB}(G)$ and the well-known fact from the computer science literature that $\alpha(G)$, $\vartheta(G)$, $\alpha^*(G)$ are the maximum values of the sum (3.19) over $\text{STAB}(G)$, $\text{TH}(G)$ and $\text{QSTAB}(G)$, respectively [GLS93, Knu94, Ros67].

### 3.2.1   The n-Cycle Inequalities

The simplest exclusivity graph for which $\beta_C < \beta_Q$ is the pentagon [CDLP13]. It can be proven by inspection that $\beta_C = 2$. The quantum bound is $\beta_Q = \sqrt{5}$, as shown by Lovász original calculation of $\vartheta(C_5)$ [Lov79]. The maximum value obtained with E-principle behaviours is $\frac{5}{2}$, which can be reached when all measurement events have probability equal to $\frac{1}{2}$.

When G is any $n$-cycle with $n$ odd, we can also prove by inspection that the classical bound is

$$\beta_C = \left\lfloor \frac{n}{2} \right\rfloor = \frac{n-1}{2}. \tag{3.25}$$

The quantum bound can also be explicitly calculated [Knu94], and we have that

$$\beta_Q = \frac{n \cos\left(\frac{\pi}{n}\right)}{1 + \cos\left(\frac{\pi}{n}\right)}, \tag{3.26}$$

which is equal to $\sqrt{5}$ for $n = 5$. The maximum obtained with E-principle behaviours is $\frac{n}{2}$, which can be reached when all measurement events have probability equal to $\frac{1}{2}$. The first five graphs in this family are shown in Fig. 3.2.

If $n$ is even, $C_n$ is a bipartite graph, and the vertices in one bipartition define a maximal independent set. The parties have the same size, and hence the classical bound is $\frac{n}{2}$. The distribution that assigns probability $\frac{1}{2}$ to all vertices achieves the bound $\beta_E$, which is then equal to $\beta_C$. The quantum bound $\beta_Q$ is sandwiched between $\beta_C$ and $\beta_E$ and hence we conclude that $\beta_Q$ is also equal to $\frac{n}{2}$.

**Fig. 3.2**  The cycles $C_n$ for $n = 5, 7, 9, 11, 13$

### 3.2.2 Vertex-Weighted Exclusivity Graphs

We can also calculate the maximum of general linear functions

$$S_\gamma = \sum_{i \in V(G)} \gamma_i p_i, \quad \gamma_i \geq 0, \tag{3.27}$$

for noncontextual, quantum and E-principle behaviours using the weighted versions of the $\alpha$, $\vartheta$ and $\alpha^*$ [Knu94, Ros67], as shown by Cabello, Severini and Winter in Ref. [CSW14].

**Definition 3.13** The *vertex-weighted exclusivity graph* $(G, \gamma)$ related to the sum (3.27) is the vertex-weighted graph defined by the exclusivity graph G and the vertex weights given by the coefficients $\gamma_i$.

**Definition 3.14** The *vertex-weighted fractional-packing number* $\alpha^*(G, \gamma)$ of $(G, \gamma)$ is defined by

$$\alpha^*(G) = \max \left\{ \sum_i \gamma_i p_i \; \middle| \; 0 \leq p_i \leq 1 \text{ and } \sum_{i \in Q} p_i \leq 1, \; Q \text{ any clique of G} \right\}. \tag{3.28}$$

**Theorem 3.7 (Cabello, Severini and Winter, 2010)** *The noncontextual, quantum and E-principle bounds for the sum (3.27) are given, respectively, by*

$$\beta_C = \alpha(G, \gamma), \quad \beta_Q = \vartheta(G, \gamma), \quad \beta_E = \alpha^*(G, \gamma), \tag{3.29}$$

*where $(G, \gamma)$ is the vertex-weighted exclusivity graph associated to the sum (3.27), $\alpha(G, \gamma)$ is the vertex-weighted independence number of $(G, \gamma)$, $\vartheta(G, \gamma)$ is the vertex-weighted Lovász number of $(G, \gamma)$ and $\alpha^*(G, \gamma)$ is the vertex-weighted fractional-packing number of $(G, \gamma)$.*

This result follows directly from the observation that $\mathscr{E}_C(G) = \text{STAB}(G)$, $\mathscr{E}_Q(G) = \text{TH}(G)$ and $\mathscr{E}_E(G) = \text{QSTAB}(G)$ and the well-known fact from the computer science literature that $\alpha(G, \gamma)$, $\vartheta(G, \gamma)$, $\alpha^*(G, \gamma)$ are the maximum values of $S_\gamma$ over STAB(G), TH(G) and QSTAB(G), respectively [GLS93, Knu94, Ros67].

### 3.2.3 Perfect Graphs

Although the characterisation of $\mathscr{E}_E(G)$ is a hard problem for general graphs, it is possible to identify the family of graphs for which there is no contextual quantum or E-principle behaviours. This is the family of *perfect graphs*.

**Definition 3.15** The *clique number* $w(G)$ of a graph G is the size of the largest clique of G.

**Definition 3.16** The *chromatic number* $\chi(G)$ of a graph G is the smallest number of colours needed to colour the vertices of G so that no two adjacent vertices share the same colour.

**Definition 3.17** A graph G is called *perfect* if

$$w(G') = \chi(G'). \tag{3.30}$$

for every induced graph G' of G.

Perfect graphs are important because several difficult problems in combinatorics can be solved in polynomial time for this special family [GLS93, Knu94, Ros67]. They were formally defined in Ref. [Ber61], where *Berge graphs* were also defined. A graph G is called a Berge graph if G or its complement does not contain a cycle $C_n$, with $n \geq 5$ odd, as an induced subgraph. Also in Ref. [Ber61], it was conjectured the equivalence of the perfect graph and Berge graph definitions. This conjecture was proved in 2002 and it is known as the *strong perfect graph theorem* [CRST06].

**Theorem 3.8 (Strong Perfect Graph Theorem)** *A graph G is perfect if, and only if, G does not contain a cycle $C_n$ or its complement $\overline{C_n}$, with $n \geq 5$ odd, as an induced subgraph.*

**Theorem 3.9** *The following are equivalent for a graph G:*

1. STAB $(G) =$ QSTAB $(G)$;
2. STAB $(G) =$ TH $(G)$;
3. TH $(G) =$ QSTAB $(G)$;
4. TH $(G)$ *is a polytope;*
5. G *is perfect.*

The proof of this result can be found in Ref. [Knu94]. As a corollary, we can state several properties of the classical, quantum and E-principle set when the exclusivity graph G is not perfect.

**Corollary 3.1** *Let G be a graph that contains a cycle $C_n$ or its complement $\overline{C_n}$, with $n \geq 5$ odd, as an induced subgraph. Then, the following hold:*

1. $\mathscr{E}_C(G) \subsetneq \mathscr{E}_Q(G)$;
2. $\mathscr{E}_Q(G) \subsetneq \mathscr{E}_E(G)$;
3. $\mathscr{E}_Q(G)$ *is not a polytope.*

This result shows that the *n*-cycle inequalities of Sect. 3.2.1 are important not only for its simplicity but also because *n*-cycles are an essential feature for a contextuality scenario to exhibit quantum contextual behaviours.

## 3.3 Coloured Multigraphs and Bell Inequalities

In this chapter, we have seen that when we remove the constraints given by a labelling of the vertices of the exclusivity graph in terms of measurement events in a specific scenario, the quantum bound for a noncontextuality inequality is *exactly* given by the vertex-weighted Lovász number of $\vartheta\,(G, \gamma)$ and the set of quantum behaviours $\mathscr{E}_Q\,(G)$ is *exactly* given by the theta body TH (G). As we have shown in Sect. 2.4, this is not the case when the scenario $(X, \mathscr{C}, O)$ is fixed. In this situation, $\vartheta\,(G, \gamma)$ is only an upper bound for $\beta_Q$ and $\mathscr{E}_Q\,(G)$ may be strictly contained in TH (G), even when $\vartheta\,(G, \gamma) = \beta_Q$.

In Ref. [RDLT+14], the authors refine the results of the previous sections, providing a graph-theoretical approach that takes into account the additional constraints of a specific labelling of the measurement events, translating the notion of separated parties into edge colours. This gives rise to an exclusivity coloured graph, which can even allow multiple edges linking the same pair of vertices.

**Definition 3.18** A *multigraph* $\Gamma$ is a graph with vertex-set V ($\Gamma$) and edge-set E ($\Gamma$) such that multiple edges between two vertices are allowed. A *vertex-weighted multigraph* $(\Gamma, \gamma)$ is a multigraph endowed with a weight assignment $\gamma \in \mathbb{R}^{|V(\Gamma)|}$ such that $\gamma_i \geq 0$ for each $i \in V\,(\Gamma)$.

We are particularly interested in *N-colour edge-coloured vertex-weighted multigraphs* $(\Gamma, \gamma)$, composed of $N$ simple vertex-weighted graphs $(G_1, \gamma), \ldots, (G_N, \gamma)$ that have the same vertex-set V ($\Gamma$) and weight assignment $\gamma$ and mutually disjoint edge-sets E ($G_1$), ..., E ($G_N$) such that E ($\Gamma$) = $\sqcup_k$E ($G_k$). Each $G_k$ is associated to a different colour $k$.

**Definition 3.19** The *exclusivity coloured graph* of an *N*-partite Bell inequality is the *N*-colour edge-coloured vertex-weighted multigraph $\Gamma$ such that each $i \in V\,(\Gamma)$ corresponds to a measurement event in the inequality and each factor $G_k$ encodes the exclusivities between party's $k$ measurement events, that is, $\{i, j\} \in E\,(G_k)$ if in measurement events $i$ and $j$ the same measurement was performed by party $k$ and different outcomes were observed.

We refer to the graph $G_k$ as the *exclusivity factor* of party $k$. This factor has several connected components, one for each of party's $k$ measurements. If a measurement has $o$ outcomes, and all of them appear in the inequality, the corresponding factor component is an $o$-partite complete graph or, equivalently, is a graph obtained from $K_o$, the $o$-vertex complete graph, by multiplying each vertex the number of times the respective outcome appears in the inequality. The coloured graph of the CHSH inequality and its exclusivity factors are shown in Fig. 3.3.

**Definition 3.20** An *orthogonal projective representation* for a graph G is an assignment of a projector $P_i$ over a real vector space to each $i \in V\,(G)$ such that $P_i P_j = 0$ whenever $\{i, j\} \notin E\,(G)$.

**Fig. 3.3** The coloured graph
$\Gamma$ for the CHSH inequality

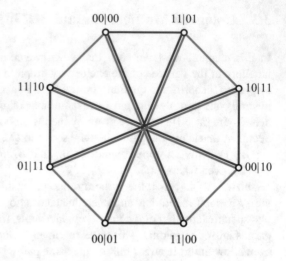

**Definition 3.21** The *coloured theta body* of a coloured graph $\Gamma$ is the set

$$\mathrm{cTH}\,(\Gamma) = \left\{ p \in \mathbb{R}^{|V(\Gamma)|} \,\middle|\, p_i = \langle \psi | \, P_i \, | \psi \rangle \right\}, \tag{3.31}$$

where $P_i = P_i^1 \otimes \ldots \otimes P_i^N$, with $\{P_i^k\}$ an orthogonal projective representation of $\overline{G_k}$ and $|\psi\rangle$ a unit vector in the appropriate vector space. The *vertex-weighted coloured Lovász number* is defined as

$$\theta\,(\Gamma, \gamma) = \sup_{p \in \mathrm{cTH}(\Gamma)} \sum_{i \in V(\Gamma)} \gamma_i \, p_i. \tag{3.32}$$

The projectors $P_i$ give an orthonormal representation for the complement of the simple graph associated to $\Gamma$. Hence, the coloured theta body is contained in the usual theta body of $\Gamma$. This also implies that

$$\theta\,(\Gamma, \gamma) \le \vartheta\,(\Gamma, \gamma). \tag{3.33}$$

Interestingly, $\theta$ and $\vartheta$ coincide for the CHSH inequality.

**Definition 3.22** The *quantum set* $\mathcal{M}_Q\,(\Gamma)$ of an $N$-colour edge-coloured vertex-weighted multigraph $\Gamma$ composed of simple graphs $G_1, \ldots, G_N$ is the set of vectors $p \in \mathbb{R}^{|V(\Gamma)|}$ whose components are given by

$$p_i = \langle \psi | \, P_i^1 \otimes \ldots \otimes P_i^N \, | \psi \rangle, \; \forall i \in V\,(\Gamma), \tag{3.34}$$

where $\{P_i^k\}$ is an orthogonal projective representation of $\overline{G_k}$ and $|\psi\rangle$ is a unit vector in the appropriate vector space.

### 3.3.1 Approximating $\theta\left(\Gamma, \gamma\right)$

The weighted Lovász number can be efficiently computed using a single semidef-
inite program [GLS81], while $\theta\left(\Gamma, \gamma\right)$ is, in general, NP-hard to approximate,
as proven in Ref. [IKM09]. Nonetheless, it is possible to adapt the famous NPA
method, introduced in Refs. [NPA07, NPA08], to outer-approximate $\mathscr{M}_Q\left(\Gamma\right)$ by
a hierarchy of SDPs. We give a brief introduction here and the reader can find a
deeper comprehensive discussion of the method in Ref. [RDLT$^+$14]. For simplicity,
we consider only bipartite Bell scenarios, but extensions to more parties are
straightforward.

Let $\left\{P_i^1\right\}$ and $\left\{P_i^2\right\}$ be orthogonal projective representations for the exclusivity
factors of parties 1 and 2, respectively. Define

$$p_{ij} = \langle\psi| P_i^1 \otimes P_j^2 |\psi\rangle, \tag{3.35}$$

where $|\psi\rangle$ is a unit vector in the appropriate vector space. With this definition, we
can write

$$\theta\left(\Gamma, \gamma\right) = \sup \sum_{i \in V(\Gamma)} \gamma_i p_{ii}, \tag{3.36}$$

where the supremum is taken over all orthogonal projective representations for $G_1$
and $\overline{G_2}$ and unit vectors $|\psi\rangle$.

In finite-dimensional Hilbert spaces, the set of vectors $P_{ij}$ defined above is known
to be the same as a set of vectors whose entries are of the form

$$P_{ij} = \langle\psi| P_i^1 P_j^2 |\psi\rangle, \tag{3.37}$$

with $\left[P_i^1, P_j^2\right] = 0$ for every $i$ and $j$. The problem of whether or not the equivalence
holds for infinite-dimensional Hilbert spaces is known as the *Tsirelson's problem*.
In the worst-case scenario, in which the maximum quantum violation of some
inequality is reached only in an infinite-dimensional Hilbert space, maximisation
over the latter will, nonetheless, give an upper bound to the maximum calculated
over the first.

We denote by $\mathscr{P} = \left\{P_i^1\right\} \cup \left\{P_j^2\right\}$ and by $\mathbf{i} = (i_1, \dots, i_n)$ a finite sequence of
numbers $i_k \in \{0, 1, \dots |\mathscr{P}|\}$ in some enumeration of the elements of $\mathscr{P}$. We define
$P_{\mathbf{i}} = P_{i_1}, \dots, P_{i_n}$ the product of the projectors $P_{i_k} \in \mathscr{P}$ associated to the elements
in sequence $\mathbf{i}$. We denote by $\mathscr{P}^{*n}$ the set of all sequences $\mathbf{i}$ with length at most $n$.

**Theorem 3.10** *Consider the matrix $M$ indexed by* $\mathbf{i}, \mathbf{j} \in \mathscr{P}^{*n}$ *with entries*

$$M_{\mathbf{i}, \mathbf{j}} := \mathrm{Tr}\left(\rho P_{\mathbf{i}} P_{\mathbf{j}}^\dagger\right) \tag{3.38}$$

*where $\rho$ is a positive-semidefinite matrix with unit trace of appropriate dimension. Then, $M$ is a positive-semidefinite matrix satisfying*

$$p_i = M_{ii, \varnothing}. \tag{3.39}$$

**Definition 3.23** We say that a model $p$ belongs to the set $\mathcal{M}_{NPA}^n(\Gamma)$ if there is a positive-semidefinite matrix $M$ indexed by $\mathbf{i}, \mathbf{j} \in \mathscr{P}^{*n}$ such that

$$p_i = M_{ii, \varnothing}. \tag{3.40}$$

Notice that Theorem 3.10 implies that

$$\mathcal{M}_Q(\Gamma) \subset \mathcal{M}_{NPA}^n(\Gamma) \tag{3.41}$$

for all $n$. Moreover, one can prove the following:

**Theorem 3.11 (Navascués, Pironio and Acín (2007))**

1. *$\mathcal{M}_{NPA}^m(\Gamma) \subset \mathcal{M}_{NPA}^n(\Gamma)$ if $m \leq n$;*
2. *$\mathcal{M}_Q(\Gamma) \subset \mathcal{M}_{NPA}^n(\Gamma)$ for all $n$;*
3. *$\mathcal{M}_Q(\Gamma) = \bigcap_n \mathcal{M}_{NPA}^n(\Gamma)$;*
4. *Each set $\mathcal{M}_{NPA}^n(\Gamma)$ can be characterised using semidefinite programming.*

The proofs of Theorems 3.10 and 3.11 can be found in Refs. [NPA08, RDLT$^+$14]. Theorem 3.11 allows for good upper approximations of maximal quantum value for Bell inequalities. The idea is to optimise the function S on the set of positive-definite matrices $M_{i,j}$ for $i, j \in \mathscr{P}^{*n}$. In Ref. [RDLT$^+$14], authors adapted this method in order to calculate upper bounds to $\theta(\Gamma, \gamma)$. By using a see-saw strategy, they could also obtain explicit projective representations for the treated examples, giving lower bounds for $\theta(\Gamma, \gamma)$. This calculates $\theta$ with precision given by the difference of these two bounds.

### 3.3.2  Pentagonal Bell Inequalities

The quantum bounds for the pentagonal Bell inequalities introduced in Example 2.8 were computed using the multigraph approach and the results coincide, up to the third digit, with the values obtained in Ref. [SBBC13]. Figure 3.4 shows the exclusivity coloured graph of the first pentagonal inequality (2.54). Figure 3.5 shows the exclusivity coloured graph for the second (2.55) and third (2.56) pentagonal inequalities.

It is interesting to note that the coloured graphs in Fig. 3.4 are isomorphic, and only the labels are different. This explains why their quantum bounds are equal. Since the pentagon in Fig. 3.5 has one more edge, coming from the double exclusivity, its quantum bound is lower than the others. Moreover, this last pentagon with one double edge can be recognised as an induced subgraph from the coloured multigraph of the CHSH inequality shown in Fig. 3.3.

**Fig. 3.4** The coloured graph
$\Gamma$ for the first pentagonal Bell
inequality

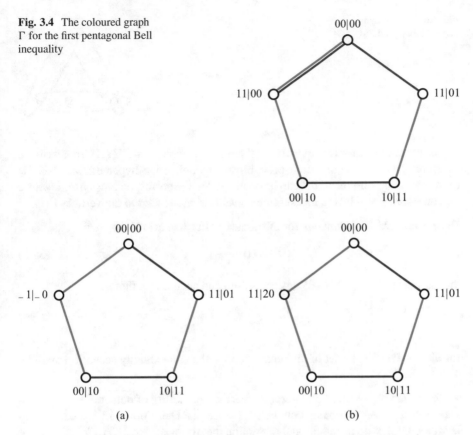

**Fig. 3.5** (a) The coloured graph $\Gamma$ for the second pentagonal Bell inequality. (b) The coloured graph $\Gamma$ for the third pentagonal Bell inequality

## 3.4 The Exclusivity-Hypergraph Approach

In Ref. [AFLS15], the authors develop a general formalism for contextuality which refines the exclusivity-graph approach, explicitly including normalisation of probabilities. The idea is to take into account not only the exclusivity relations among the measurement events but also to include the information of which measurement gave rise to each specific exclusivity. To this end, a contextuality scenario in this formalism will be associated to a hypergraph whose vertices represent the measurement events in the scenario and the hyperegdes represent the measurements.

**Definition 3.24** A *contextuality scenario* in the exclusivity-hypergraph approach is defined by a hypergraph H with vertex-set V (H) and edge-set E (H) such that

$$\bigcup_{\epsilon \in E(H)} \epsilon = V(H). \tag{3.42}$$

**Fig. 3.6** The triangle
contextuality scenario

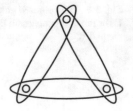

As in the exclusivity approach of Sect. 3.1, each $i \in$ V (H) represents a
measurement event $e_i$ in some probabilistic model. Each hyperedge $\epsilon \in$ E (H)
represents a measurement that can be conducted in the corresponding system, whose
outcomes are given by the measurement events $e_i$ associated to the vertices $i \in \epsilon$.

**Definition 3.25** A *behaviour* for the contextuality scenario H is a map

$$p : V (H) \longrightarrow [0, 1] \tag{3.43}$$

that assigns to each $i \in$ V (H) a probability $p_i$ in such a way that

$$\sum_{i \in \epsilon} p_i = 1 \tag{3.44}$$

for all $\epsilon \in$ E (H). The set of all behaviours for the contextuality scenario H will be
denoted by $\mathscr{G}$ (H) .

The number $p_i$ should be interpreted as the probability of outcome $e_i$ given that
a measurement $\epsilon \ni i$ is being conducted. Notice that Definition 3.25 is analogous to
Definition 3.2, with the additional restriction that for every $\epsilon \in$ E (H), $\sum_{i \in \epsilon} p_i$ must
be *exactly* one. This guarantees that for each $\epsilon$ the vector $(p_i)_{i \in \epsilon}$ is a probability
vector. This is in contrast with what has been done in Sect. 3.1, where there might
be some outcomes of the measurements under consideration that do not appear
in the exclusivity graph. For example, the assignment $p_i = 0$ for all $i$ defines a
valid behaviour in the exclusivity-graph approach, but it does not in the exclusivity-
hypergraph approach.

By Definition 3.25, *measurement noncontextuality* is assumed for all behaviours:
the probability associated to any outcome in V (H) does not depend on the particular
measurement in which the outcome occurs.

Although the exclusivity graph does not define uniquely the contextuality
scenario in the exclusivity-hypergraph approach, it will also be important for the
characterisation of the various convex sets associated to a contextuality scenario.

**Definition 3.26** The *exclusivity graph* G of the contextuality scenario H is the 2-
skeleton of the hypergraph H, that is, V (G) = V (H) and $\{i, j\} \in$ E (G) if, and only
if, there is a hyperedge $\epsilon \in$ E (H) such that $i, j \in \epsilon$.

*Example 3.1* Figure 3.6 shows the *triangle scenario*, where we have three measurements with two outcomes each.

**Fig. 3.7** A contextuality scenario with no behaviour

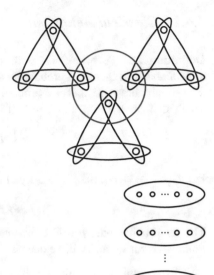

**Fig. 3.8** The contextuality scenario with $k$ measurements with $m$ different outcomes each

The normalisation condition implies the system of linear equations

$$\begin{cases} p_1 + p_2 - 1 \\ p_2 + p_3 = 1 \\ p_1 + p_3 = 1, \end{cases} \tag{3.45}$$

which has the unique solution

$$p_1 = p_2 = p_3 = \frac{1}{2}. \tag{3.46}$$

Hence, the contextuality scenario given by this hypergraph has a unique behaviour, defined by Eq. (3.46).

*Example 3.2* Figure 3.7 shows a contextuality scenario with $\mathscr{G}(H) = \varnothing$. In fact, the vertices in the central (blue) hyperedge must have probability equal to $\frac{1}{2}$, as proved in the previous example, but this is a contradiction since these probabilities must add to 1.

*Example 3.3* Figure 3.8 shows a scenario with $k$ $m$-outcome measurements, such that no two measurements have common outcomes. We will see later that this scenario will be useful to describe the Bell scenario $(N, k, m)$ in terms of the Foulis-Randall product, to be defined in Sect. 4.5.

### 3.4.1  Classical Behaviours

**Definition 3.27** A behaviour $p$ in a contextuality scenario H is *classical* or *noncontextual* if there is a probability space $(\Omega, \Sigma, \mu)$, where $\Omega$ is a sample space, $\Sigma$ a $\sigma$-algebra in $\Omega$ and $\mu$ a probability measure in $\Sigma$, and for each $i \in V(H)$ a set $A_i \in \Sigma$ such that $A_i \cap A_j = \varnothing$ if $i, j$ belong to some $\epsilon \in E(H)$ and

$$\bigcup_{i \in \epsilon} A_i = \Omega. \tag{3.47}$$

For each $i \in V(H)$, the probability of the measurement event associated to $i$ is

$$p_i = \mu(A_i). \tag{3.48}$$

The set of behaviours $p \in \mathbb{R}^{|V(H)|}$ obtained in this way is called the *classical set* or *noncontextual set* and will be denoted by $\mathcal{H}_C(H)$.

This is similar to Definition 3.3, with the extra assumption that probabilities should be normalised for every hyperedge $\epsilon \in E(H)$, guaranteed by Eq. (3.47). Hence, we have that

$$\mathcal{H}_C(H) \subset \mathcal{E}_C(G), \tag{3.49}$$

where G is the exclusivity graph associated to the scenario H.

Definition 3.27 is not the one used in Ref. [AFLS15], but Theorem 3.12 shows that both ways of defining the classical set are equivalent.

**Definition 3.28** A behaviour $p$ in a contextuality scenario H is *deterministic* if $p_i \in \{0, 1\}$ for every $i \in V(H)$.

**Theorem 3.12** *A behaviour $p$ in a contextuality scenario H is classical if, and only if, it is a convex combination of deterministic behaviours.*

The proof follows the same lines as the proof of Theorem 3.1. A consequence of this result is the following:

**Corollary 3.2** *Let $p$ be a behaviour for H obtained with some probabilistic model. Then, $p$ can also be obtained with a completion of this probabilistic model satisfying measurement noncontextuality if, and only if, $p$ is classical.*

In fact, the extremal behaviours in any completion satisfying measurement noncontextuality must be exactly the deterministic behaviours for H. This, in turn, gives a characterisation of the classical set in terms of the stable set polytope of the exclusivity graph of the scenario.

**Theorem 3.13** *Let G be a graph. Then, we have that*

$$\mathrm{STAB}(G) = \left\{ p \in \mathbb{R}^{|V(G)|} \,\middle|\, \alpha^*(\overline{G}, p) \leq 1 \right\}, \tag{3.50}$$

*where $\alpha^*(\overline{G}, p)$ is the vertex-weighted fractional-packing number of $(\overline{G}, p)$.*

The proof of this result can be found in Ref. [Knu94].

**Lemma 3.1** *If p is a behaviour for* H, *then*

$$\alpha\left(\overline{G}, p\right) \geq 1. \tag{3.51}$$

*Proof* If $p$ is a behaviour for H, then $p$ satisfies conditions (3.44). Take any $\epsilon \in$ E (H) and $Q \subset$ V (H) the corresponding clique in G. The set $Q$ is an independent set in $\overline{G}$ and the $q^*$ vector defined by

$$q_i^* = \begin{cases} 1, & \text{if } i \in Q \\ 0, & \text{if } i \notin Q \end{cases} \tag{3.52}$$

belongs to STAB $\left(\overline{G}\right)$. Hence, we have

$$\alpha\left(\overline{G}, p\right) = \max_{q \in \text{STAB}(\overline{G})} \sum_{i \in V(G)} p_i q_i \geq \sum_{i \in V(G)} p_i q_i^* = \sum_{i \in Q} p_i = 1. \tag{3.53}$$

$\square$

**Theorem 3.14** *A behaviour p in a contextuality scenario* H *is classical if, and only if,*

$$\alpha^*\left(\overline{G}, p\right) = 1. \tag{3.54}$$

*Proof* Since $\mathcal{E}_C$ (G) $=$ STAB (G), Theorem 3.13 implies that membership in $\mathcal{E}_C$ (G) is equivalent to

$$\alpha^*\left(\overline{G}, p\right) \leq 1. \tag{3.55}$$

As proven in Lemma 3.1, the normalisation condition implies that

$$\alpha^*\left(\overline{G}, p\right) \geq \alpha\left(\overline{G}, p\right) \geq 1. \tag{3.56}$$

The proof follows from the fact that the elements of $\mathcal{H}_C$ (H) are exactly the elements of $\mathcal{E}_C$ (G) for which the normalisation condition holds.          $\square$

### 3.4.2   Quantum Behaviours

**Definition 3.29** A behaviour $p$ in a contextuality scenario H is *quantum* if there is a density matrix $\rho$ acting on a Hilbert space $\mathfrak{H}$ and for each $i \in$ V (H) a projector $P_i$ acting on $\mathfrak{H}$ such that $P_i$ and $P_j$ are orthogonal if $\{i, j\} \subset \epsilon$ for some $\epsilon \in$ E (H),

$$\sum_{i \in \epsilon} P_i = I \tag{3.57}$$

for every $\epsilon \in E(H)$, and such that for each $i$ the probability of measurement event associated to $i$ is

$$p_i = \text{Tr}(P_i \rho). \tag{3.58}$$

The set of all quantum behaviours is called the *quantum set* and will be denoted by $\mathcal{H}_Q(H)$.

**Theorem 3.15** *The quantum set $\mathcal{H}_Q(H)$ is convex and $\mathcal{H}_C(H) \subset \mathcal{H}_Q(H)$.*

The proof is analogous to the ones presented in Sects. 2.2.2 and 3.1.2.

Differently to what happens in the exclusivity-graph approach, in the exclusivity-hypergraph approach the quantum set cannot be related to any graph invariant of the exclusivity graph G. In fact, one can prove the following result:

**Theorem 3.16** *There exist two contextuality scenarios H and H' with the same exclusivity graph $G = G'$ and a behaviour p that is quantum for H but not for H'.*

The proof of this result can be found in Ref. [AFLS15]. This result shows that the restrictions enforced by the hypergraph H and the requirement of normalisation of probabilities are much more intricate than the ones imposed by the exclusivity graph G alone.

In Appendix A, we will discuss in detail several state-independent proofs of the impossibility of noncontextual completions of quantum theory. Interestingly, we can rewrite these proofs elegantly in terms of the formalism just presented.

**Theorem 3.17 (Bell-Kochen-Specker)** *There are contextuality scenarios such that $\mathcal{H}_C(H) = \varnothing$ and $\mathcal{H}_Q(H) \neq \varnothing$.*

Such a contextuality scenario automatically provides a state-independent proof of the Bell-Kochen-Specker theorem, since the fact that $\mathcal{H}_Q(H) \neq \varnothing$ implies that there is an assignment of projectors in some Hilbert space to the vertices of H satisfying the conditions of Definition 3.29. Any state in this Hilbert space provides a contextual behaviour, since $\mathcal{H}_C(H) = \varnothing$. Any of the so-called Kochen-Specker sets we will present in Appendix A defines such a scenario.

### 3.4.3   Approximating the Quantum Set

Let $p$ be a quantum behaviour such that

$$p_i = \text{Tr}(\rho P_i). \tag{3.59}$$

We denote by $\mathbf{i} = (i_1, \ldots, i_n)$ a finite sequence of vertices $i_k \in V(H)$ and by $P_\mathbf{i} = P_{i_1} \ldots P_{i_n}$ the product of the projectors associated to the vertices in sequence $\mathbf{i}$. We denote by V (H)$^{*n}$ the set of all sequences $\mathbf{i}$ with length at most $n$. If $\mathbf{i} = (i_1, \ldots, i_k)$ and $l \in$ V (H), we denote by $\mathbf{i}l = (i_1, \ldots, i_k, l)$ the juxtaposition of $\mathbf{i}$ and $l$.

**Theorem 3.18** *The matrix M indexed by* $\mathbf{i}, \mathbf{j} \in$ V (H)$^{*n}$ *with entries*

$$M_{\mathbf{i}, \mathbf{j}} := \mathrm{Tr} \left( \rho P_{\mathbf{i}} P_{\mathbf{j}}^{\dagger} \right) \tag{3.60}$$

*is a positive-semidefinite matrix satisfying*

1. *$M_{\varnothing, \varnothing} = 1$;*
2. *For every $\epsilon \in$ E (H)*

$$\sum_{l \in \epsilon} M_{\mathbf{i}, \mathbf{j}l} = M_{\mathbf{i}, \mathbf{j}}; \tag{3.61}$$

3. *If $\mathbf{i} = (i_1, \ldots, i_k)$ and $\mathbf{j} = (j_1, \ldots, j_m)$ are exclusive, then*

$$M_{\mathbf{i}, \mathbf{j}} = 0. \tag{3.62}$$

Notice that

$$p_i = M_{i, \varnothing}. \tag{3.63}$$

**Definition 3.30** We say that a model $p$ belongs to the set $\mathcal{H}^n_{AFLS}$ (H) if there is a positive-semidefinite matrix $M$ indexed by $\mathbf{i}, \mathbf{j} \in$ V (H)$^{*n}$ satisfying all conditions listed in Theorem 3.18 such that

$$p_i = M_{i, \varnothing}. \tag{3.64}$$

**Theorem 3.19**

1. $\mathcal{H}^m_{AFLS}$ (H) $\subset \mathcal{H}^n_{AFLS}$ (H) *if $m \leq n$;*
2. $\mathcal{H}_Q$ (H) $\subset \mathcal{H}^n_{AFLS}$ (H) *for all $n$;*
3. $\mathcal{H}_Q$ (H) $= \bigcap_n \mathcal{H}^n_{AFLS}$ (H)*;*
4. *Each set $\mathcal{H}^n_{AFLS}$ (H) can be characterised using semidefinite programming.*

The proofs of Theorems 3.18 and 3.19 can be found in Ref. [AFLS15]. The set $\mathcal{H}^1_{AFLS}$ (H) is of special interest for us. It can be characterised in terms of orthonormal representations of the exclusivity graph G.

**Theorem 3.20** *A behaviour $p$ belongs to $\mathcal{H}^1_{AFLS}$ (H) if, and only if, there is a Hilbert space $\mathfrak{H}$, a unit vector $|\psi\rangle \in \mathfrak{H}$ and a projector $P_i$ acting on $\mathfrak{H}$ for every $i \in$ V (H) such that $P_i$ and $P_j$ are orthogonal if $\{i, j\} \in$ E (G) and*

$$p_i = \langle \psi | P_i |\psi \rangle \tag{3.65}$$

*for all $i \in$ V (H).*

Theorem 3.20 implies that $\mathcal{H}^1_{AFLS}$ (H) $\subset \mathcal{E}_Q$ (G) and that it consists exactly of the elements of $\mathcal{E}_Q$ (G) such that $\sum_{i \in \epsilon} p_i = 1$ for every $\epsilon \in$ E (H).

**Corollary 3.3** *A behaviour p belongs to* $\mathscr{H}^1_{AFLS}$ (H) *if, and only if,*

$$\vartheta\left(\overline{G}, p\right) = 1, \tag{3.66}$$

*where* $\vartheta\left(\overline{G}, p\right)$ *is the vertex-weighted Lovász number of* $\left(\overline{G}, p\right)$.

*Proof* The fact that $\mathscr{H}^1_{AFLS}$ (H) $\subset \mathscr{E}_Q$ (G) implies that $\vartheta\left(\overline{G}, p\right) \leq 1$. Normalisation of probabilities for every $\epsilon \in E$ (H) implies that

$$\vartheta\left(\overline{G}, p\right) \geq \alpha\left(\overline{G}, p\right) \geq 1. \tag{3.67}$$

□

### 3.4.4   E-Principle Behaviours

In the exclusivity-hypergraph approach, the E-principle, referred to as *consistent exclusivity* in Ref. [AFLS15], also imposes nontrivial constraints on the behaviours in $\mathscr{G}$ (H).

**Definition 3.31** Given a contextuality scenario H, a behaviour $p \in \mathscr{G}$ (H) satisfies the *Exclusivity Principle* if

$$\sum_{i \in Q} p_i \leq 1 \tag{3.68}$$

for every clique $Q \in E$ (G). The set of all behaviours satisfying the E-principle will be denoted by $\mathscr{H}_E$ (H).

**Theorem 3.21** *Given a contextuality scenario* H, $\mathscr{H}_Q$ (H) $\subset \mathscr{H}_E$ (H).

The proof follows the same lines as the proof of Theorem 3.5.

There are scenarios for which $\mathscr{H}_E$ (H) $\subsetneq \mathscr{G}$ (H). For example, the triangle scenario of Fig. 3.6 has a unique behaviour $p$ for which the sum $\sum_i p_i = \frac{3}{2}$ for the unique clique in the exclusivity graph G.

Also in the exclusivity-hypergraph approach, the E-principle set can be characterised in terms of the clique-constrained stable set polytope of the exclusivity graph.

**Theorem 3.22** *Let* G *be a graph. Then, we have that*

$$\text{QSTAB (G)} = \left\{ p \in \mathbb{R}^{|V(G)|} \,\middle|\, \alpha\left(\overline{G}, p\right) \leq 1 \right\}, \tag{3.69}$$

*where* $\alpha\left(\overline{G}, p\right)$ *is the vertex-weighted independence number of* $\left(\overline{G}, p\right)$.

The proof of this result can be found in Ref. [Knu94].

**Theorem 3.23** *A behaviour p belongs to $\mathcal{H}_E$ (H) if, and only if,*

$$\alpha\left(\overline{G}, p\right) = 1, \tag{3.70}$$

*where G is the exclusivity graph of the scenario H.*

*Proof* Since $\mathcal{E}_E$ (G) = QSTAB (G), Theorem 3.22 implies that membership in $\mathcal{E}_E$ (G) is equivalent to

$$\alpha\left(\overline{G}, p\right) \leq 1. \tag{3.71}$$

The normalisation condition $\sum_{i \in \epsilon} p_i = 1$ for every $\epsilon \in E$ (H) implies that

$$\alpha\left(\overline{G}, p\right) \geq 1. \tag{3.72}$$

□

### 3.4.5 Perfect Graphs

The fact that the classical and E-principle sets are related to STAB (G) and QSTAB (G), respectively, and the results of Sect. 3.2.3 imply that there is no E-principle contextual behaviour if G is perfect.

**Theorem 3.24** *If G is perfect, then $\mathcal{H}_C$ (H) = $\mathcal{H}_Q$ (H) = $\mathcal{H}_E$ (H), although $\mathcal{G}$ (H) can still be bigger.*

As a corollary of the strong perfect graph Theorem 3.8, we have:

**Corollary 3.4** *If neither G nor $\overline{G}$ contains an odd cycle of length n $\geq$ 5 as an induced subgraph, then $\mathcal{H}_C$ (H) = $\mathcal{H}_Q$ (H) = $\mathcal{H}_E$ (H).*

In the exclusivity-hypergraph approach, the converse does not hold. There are contextuality scenarios for which $\mathcal{G}$ (H) = $\mathcal{H}_C$ (H) although G is not perfect. An example of such a scenario can be found in Ref. [AFLS15].

## 3.5 Final Remarks

Another perspective to contextuality is given by the exclusivity-graph approach. We start with the exclusivity graph G, where each vertex $i$ represents a measurement event. If $\{i, j\} \in E$ (G), the measurement events $i$ and $j$ are exclusive, that is, there is at least one common measurement with distinct outcomes described by $i$ and $j$. The main difference between this approach and the compatibility graph approach is that in this case we make no restriction in the compatibility scenario leading to the exclusivity structure of the events. This offers a wider spectrum of interesting phenomena.

As we have seen, the same measurement event may correspond to an outcome of several different measurements. In this new perspective, the noncontextuality assumption implies that whenever a measurement event corresponds to outcomes of different measurements $M_1, M_2, \ldots, M_n$, a noncontextual completion assigns the outcome corresponding to this event to some $M_i$ if and only if it does for all other $M_j$.

The set of noncontextual behaviour is once more a polytope, and the quantum set is generally larger. The classical set can be described by a finite set of noncontextuality inequalities, violated by quantum distributions in many situations.

The noncontextual, quantum and generalised bounds for the noncontextuality inequality can be found using the exclusivity graph of the scenario. The classical bound is equal to the independence number of the exclusivity graph and the quantum bound is *equal* to the Lovász number of this graph. In this case, we have an equality between the quantum bound and the Lovász number because we do not have extra restrictions imposed by a specific compatibility structure.

Sections 3.3 and 3.4 present refinements of the exclusivity-graph approach, where other structures are taken into account when looking at the exclusivities between the measurement events. In the first case, the exclusivities between the measurement events in a Bell scenario are distinguished to take into account from which of the parties the exclusivity comes from. In the second case, besides exclusivity, normalisation of measurements is also a key concept. In these cases, the Lovász number is, in general, only an upper bound and the quantum set is hard to characterise. This illustrates how quantum behaviours can be more complicated when extra restrictions are imposed on the measurement events and gives a hint on the importance of focusing on exclusivities only.

The most general distributions we consider have to satisfy the Exclusivity principle and for this kind of distribution the bound is equal to the fractional-packing number of the exclusivity graph. This principle will be used in Chap. 4 in our attempt to understand why quantum theory is not *more noncontextual* than it is.

# Chapter 4
# The Exclusivity Principle
# and Its Consequences

The mathematical formulation of quantum theory is almost one century old and since its development a number of brilliant scientists around the world have built a quite good knowledge about it, both on theoretical aspects and experimental control of quantum systems. Physicists are capable nowadays of making remarkably accurate descriptions of molecular structure of matter, high-energy particle collisions, semiconductor behaviour, spectral emissions, and much more [Bal13], as well as using all this knowledge to built new devices and protocols for computation and information processing. This certainly has a great impact on the development of current technology [ABB+17].

From the practical point of view, we may say that physicists have a good relationship with quantum theory. But, just as Einstein, Podolsky and Rosen (EPR) in 1935 [EPR35, Sta], you can get in serious trouble when you try to understand the *meaning* of the mathematical objects, specially if you try to reconcile the theory with some aspects of "classical reality" that seem natural to our intuition, based on our experience with macroscopic systems.

This situation led many people to adopt the way of thinking known as *Copenhagen interpretation*. According to this line of thought, the weirdness of quantum theory reflects fundamental limits on what can be known about nature and we just have to accept it. Quantum theory should not be understood but seen as a tool to get practical results. As famously phrased by David Mermin, physicist should "shut up and calculate" [Mer89].

Not everyone is happy with this interpretation, including Mermin himself [Mer14]. Physics is not just about getting practical results, it is also about *understanding* how nature behaves. Since the Einstein-Bohr debate [Sta], many have tried to understand (or question, like EPR) the abstract formulation of quantum theory from more compelling physical arguments. This is one of the most fascinating scientific challenges in recent times: deriving quantum theory from simple physical principles.

© The Author(s), under exclusive licence to Springer Nature Switzerland AG 2018　　　75
B. Amaral, M. Terra Cunha, *On Graph Approaches to Contextuality
and their Role in Quantum Theory*, SpringerBriefs in Mathematics,
https://doi.org/10.1007/978-3-319-93827-1_4

The starting point is assuming general probabilistic theories allowing for probability distributions that are more general than those that arise in quantum theory, and the goal is to find principles that pick out quantum theory from this landscape of possible theories. There are diverse ideas on how to do this, and at least three different approaches to the problem stand out.

The first one consists of reconstructing quantum theory as a purely operational probabilistic theory that follows from some sets of axioms [Har01, Har11, MM11, CDP11, AG16]. Imposing a small number of physical principles, these reconstructions manage to prove that the only consistent probabilistic theory is quantum. Although successful in many aspects, this approach does not resolve the issue completely, specially because some of the principles imposed on the probabilistic theories are not so natural as one would expect. This "unsatisfaction" is very well synthesised by Chris Fuchs [FS16]:

> There is no doubt that this is invaluable work, particularly for our understanding of the intricate connections between so many quantum information protocols. But to me, it seems to miss the mark for an ultimate understanding of quantum theory; I am left hungry. I still want to know what strange property of matter forces this formalism upon our information accounting. I would like to see an axiomatic system that goes for the weirdest part of quantum theory.

The second approach to the problem goes in this direction. Instead of trying to reconstruct quantum theory, the idea is to understand what physical principles explain one of its weirdest parts: nonlocality. Many different principles have been proposed, such as *non-triviality of communication complexity* [vD12, BBL+06], *no advantage for nonlocal computation* [LPSW07], *information causality* [PPK+09], *macroscopic locality* [NW09] and *local orthogonality* [FSA+13]. These principles give an important steps towards a deeper comprehension of quantum theory, but none of them was able to explain the entire set of quantum behaviours for general scenarios.

The third approach consists in identifying principles that explain the set of quantum contextual behaviours without restrictions imposed by a specific experimental scenario. The belief that identifying the physical principle responsible for quantum contextuality can be more successful than previous approaches is based on two observations. On one hand, when focusing on quantum contextuality, we are just considering a natural extension of quantum nonlocality which is free of certain restrictions (composite systems, space-like separated tests with multiple observers, entangled states). With these restrictions, one cannot explain all the features of quantum theory, living apart some fundamental examples like a three-level system, although they are crucial for many important applications, specially in communication protocols (see, for example, Refs. [Wikc, HHHH09, BBC+93] and other references therein), and played an important role in the historical debate on whether or not quantum theory is a complete theory.[1]

---

[1]In a sense, the nonlocality approach considers information-theoretical tasks like communication from two or more parties as starting points, while the contextuality approach tends to consider that the basic information-theoretical task is the description of a system or experiment.

On the other hand, it is based on the observation that, while calculating the maximum value of quantum behaviours for nonlocality scenarios is a mathematically complex problem (see [PV10] to see how complex it is to get the quantum maximum for a simple inequality like $I_{3322}$), calculating the maximum contextual value of quantum behaviours for an *arbitrary* scenario characterised by its exclusivity graph is simple: as we proved in Sect. 3.2, the maximum quantum contextuality is given by the Lovász number of its exclusivity graph, which is the solution of a semidefinite program [Lov95]. Indeed, from the exclusivity-graph approach perspective, the difficulties in characterising quantum nonlocal behaviours are due to the mathematical difficulties associated to the extra constraints resulting from enforcing a particular labeling of the events of an exclusivity structure in terms of parties, local settings, and outcomes [SBBC13], rather than a fundamental difficulty related to the principles of quantum theory.

Within this line of research, the most promising candidate for being *the* fundamental principle of quantum contextuality is the E-Principle, which states that the sum of the probabilities of a set of pairwise exclusive events cannot exceed 1, as we saw in Sect. 3.1.

The E-principle was suggested by the works of Specker [Spe60] and Wright [Wri78] and used in Ref. [CSW10] as an upper bound for quantum contextuality. However, its fundamental importance for quantum theory was conjectured long before [Spe]. It was promoted to a possible fundamental principle by the observation that it explains the maximum quantum violation of the simplest noncontextuality inequality, as we will see in Sect. 4.2. It also explains the quantum maxima for many other inequalities and rules out many nonlocal boxes in important Bell scenarios. The E-principle, when applied only to Bell scenarios, is also called *local orthogonality* [FSA+13]. However, *with this extra restriction*, the E-principle cannot rule out the non-quantum behaviours that belong to the *almost quantum set* that in general contains the quantum set properly [FSA+13, NGHA15].

By itself, the E-principle singles out the maximum quantum value for some Bell and noncontextuality inequalities [Cab13b]. According to the results of Sect. 3.2, this happens whenever $\vartheta(G) = \alpha^*(G)$. We can get better bounds if we apply the E-principle to more sophisticated scenarios. When applied to the OR product of two copies of the exclusivity graph, which physically may be seen as two independent realisations of the same experiment,[2] it was shown in Ref. [Cab13b] that the E-principle singles out the maximum quantum value for experiments whose exclusivity graphs are vertex-transitive and self-complementary, which include the KCBS inequality presented in Sect. 3.2.1. Moreover, either applied to two copies of the exclusivity graph of the CHSH inequality or of the simpler pentagonal Bell inequality (2.54), the E-principle excludes the so-called PR boxes [PR94], and provides an upper bound to the maximum violation of the CHSH inequality which is close to the Tsirelson bound [FSA+13, Cab13b]. In addition, when applied to the OR product of an infinite number of copies, there is strong evidence that the

---

[2]Or even two independently samples of data from the same experiment.

E-principle singles out the maximum quantum violation of the noncontextuality inequalities whose exclusivity graph is the complement of odd cycles on $n \geq 7$ vertices [CDLP13]. Indeed, it might be also the case that, when applied to an infinite number of copies, the E-principle singles out the Tsirelson bound of the CHSH inequality [FSA+13, Cab13b].

Another evidence of the strength of the E-principle was found by Yan in Ref. [Yan13]. By exploiting lemma 1 in Ref. [Lov79], Yan has proven that *if all behaviours predicted by quantum theory for an experiment with exclusivity graph* G *are reachable in nature,* then the E-principle singles out the *maximum* value of the behaviours produced by an experiment whose exclusivity graph is the complement of G.

Stronger consequences of the E-principle are presented in Ref. [ATC14]. The main result states that the E-principle singles out the *entire set of quantum behaviours* associated to any exclusivity graph *assuming* the set of quantum behaviours for the complementary graph. Moreover, for self-complementary graphs, the E-principle, *by itself* (i.e., without further assumptions), excludes any set of behaviours strictly larger than the quantum set. Finally, for vertex-transitive graphs, the E-principle singles out the maximum value of a specific noncontextuality inequality attained by quantum behaviours, assuming only the quantum maximum for the complementary graph. These results show that the E-principle goes beyond any other proposed principle towards the objective of singling out the set of quantum behaviours in the exclusivity-graph approach.

It is important to stress that the E-principle does not explain the quantum set in the compatibility-hypergraph nor in the exclusivity-hypergraph approach. This should not be a surprise since the E-principle rests on the exclusiveness relations of the events, which are not sufficient to describe quantum sets coming from measurement scenarios, where we enforce several extra restrictions on the measurement events. Nonetheless, the E-principle strongly restricts the set of behaviours in those approaches as well, as we will see in Sect. 4.5. For example, the E-principle, applied to a sophisticated combination of scenarios, explains the quantum bound for the CGLMP Bell inequalities [CGL+02], which includes the CHSH inequality as a special case.

## 4.1   The Exclusivity Principle

To start with, let us briefly review some of the definitions and concepts introduced in Sect. 3.2. We start with an exclusivity graph G. Each vertex $i \in V(G)$ corresponds to a measurement event in a probabilistic model and two vertices are connected by an edge if they are exclusive, that is, if they are associated to two different outcomes of some measurement. For a given state of the system, there is a probability $p_i$ associated to each vertex $i \in V(G)$. We collect all these probabilities in the behaviour $p \in \mathbb{R}^{|V(G)|}$.

In Chap. 3, we have seen that the E-principle constrains the set of behaviours for an exclusivity graph G, but in general we still have that

$$\mathscr{E}_Q(G) \subsetneq \mathscr{E}_E(G). \tag{4.1}$$

This implies that the E-principle, applied to the behaviours $p$ alone, is not capable of singling out the quantum set for a general exclusivity graph. Nonetheless, the E-principle displays *activation effects*: a behaviour $p \in \mathscr{E}_E(G)$ does not necessarily satisfy the E-principle when combined with some other behaviour $q \in \mathscr{E}_Q(G')$, where $G'$ can be any other exclusivity graph.

**Definition 4.1** The *OR product* of two graphs G and G′ is the graph $G \otimes G'$ such that $V(G \otimes G') = V(G) \times V(G')$ and $\{\{i, i'\}, \{j, j'\}\} \in E(G \otimes G')$ if either $\{i, j\} \in E(G)$ or $\{i', j'\} \in E(G')$. We denote the OR product of $n$ copies of G with itself by $G^{\otimes n}$.

Among many products of graphs, this is a natural choice when considering two *exclusivity* graphs: a pair of collective events is exclusive if any of its components is exclusive.

**Theorem 4.1** *Let $\{e_i\}$ be a set of measurement events with exclusivity graph G and $\{e'_j\}$ be a set of measurement events with exclusivity graph G′, such that $e_i$ and $e'_j$ are independent. Define the event $f_{ij} = e_i \wedge e'_j$ which is true if and only if both $e_i$ and $e'_j$ are true. Then the exclusivity graph associated to the set of events $\{f_{ij}\}$ is $G \otimes G'$.*

*Proof* To see that this is indeed the case, we just have to notice that $f_{ij}$ and $f_{kl}$ are exclusive either if $e_i$ and $e_k$ are exclusive or if $e'_j$ and $e'_k$ are exclusive. □

**Corollary 4.1** *Let $\{e_i\}$ be a set of n measurement events with exclusivity graph G and $\{e'_j\}$ be a set of n measurement events with exclusivity graph $\overline{G}$, such that $e_i$ and $e'_j$ are independent. Define the event $f_{ii} = e_i \wedge e'_i$ which is true if and only if both $e_i$ and $e'_i$ are true. Then the exclusivity graph associated to the set of events $\{f_{ii}\}$ is the complete graph on n vertices $K_n$.*

*Proof* The fact that the exclusivity graph of $\{e_i\}$ is G and the exclusivity graph of $\{e'_j\}$ is $\overline{G}$ implies that $f_{ii} = e_i \wedge e'_i$ and $f_{jj} = e_j \wedge e'_j$ are always exclusive, since either $e_i$ and $e_j$ are exclusive or $e'_i$ and $e'_j$ are exclusive. This in turn implies that the induced subgraph of $G \otimes G'$ corresponding to the events $f_{ii}$ is $K_n$. □

To each pair of behaviours $p$ and $p'$ for G and G′, respectively, we associate the behaviour $p \otimes p'$ for $V(G \otimes G')$ defined as

$$p \otimes p'(i, j) = p(i) p'(j). \tag{4.2}$$

This is the behaviour associated to the independent realisation of the experiments associated to G and G′ in which the probabilities for the measurement events in G are given by $p$ and probabilities for the measurement events in G′ are given by $p'$. The product of $n$ copies of a behaviour $p$ with itself, as defined by Eq. (4.2), will be denoted by $p^{\otimes n}$.

It is reasonable to assume that if the E-principle is a physical law that we should impose on the behaviours in a contextuality scenario, then the independent realisation of arbitrary behaviours should also satisfy this principle in the large scenario. Hence, we consider also the following refinements of the E-principle:

**Definition 4.2** We denote by $\mathscr{E}_E^n$ (G) the set of behaviours $p \in \mathscr{E}_E$ (G) such that

$$p^{\otimes n} \in \mathscr{E}_E \left( G^{\otimes n} \right). \tag{4.3}$$

Furthermore, we define

$$\mathscr{E}_E^\infty (G) = \bigcap_{n \in \mathbb{N}} \mathscr{E}_E^n (G). \tag{4.4}$$

**Definition 4.3 (Complete Exclusiveness)** We denote by $\tilde{\mathscr{E}}_E$ (G) the set of behaviours $p \in \mathscr{E}_E$ (G) such that

$$p \otimes q \in \mathscr{E}_E \left( G \otimes G' \right) \tag{4.5}$$

for every behaviour $q \in \mathscr{E}_Q$ (G′). and every contextuality scenario G′.

In this chapter we will see how the activation effects of the E-principle constrains even more the set of all possible behaviours. The ultimate goal would be to prove that the E-principle is the reason why quantum theory is not *more contextual*, or, in a sense, that quantum theory is the most contextual among the reasonable theories, and that this principle explains the quantum maximum achieved for every noncontextuality inequality. It is not clear what happens in general, but for a special class of inequalities (or graphs) many results supporting this conjecture have been found. We will apply the E-principle for sums of the type:

$$S = \sum_{i \in V(G)} p_i, \tag{4.6}$$

that is, we set $\gamma_i = 1$ for all $i$ in Definition 3.11. For noncontextual behaviours we know that

$$S \overset{\text{NC}}{\leq} \alpha (G), \tag{4.7}$$

while for quantum behaviours we have

$$S \overset{Q}{\leq} \vartheta\,(G)\,, \tag{4.8}$$

where $\vartheta\,(G)$ is the Lovász number of G.

The first question is if the E-principle is capable of explaining the quantum bound $\vartheta\,(G)$. For many different cases with special importance for the study of contextuality, this is indeed the case. A much more ambitious question is if this principle is enough to single out the set of quantum behaviours and not just the quantum maximum. Again, we are able to exhibit an important family of graphs for which the answer is affirmative.

## 4.2   The Pentagon

The E-principle singles out the quantum maximum for the simplest noncontextuality inequality.

**Theorem 4.2 (Cabello, 2013)** *For G = C5, the maximum value for S allowed by theories satisfying the E-principle is $\sqrt{5}$, which is also the maximum achieved by quantum behaviours.*

*Proof* Let $\{e_i\}$ and $\{e_i'\}$ be two sets of 5 events with exclusivity graphs as shown in Fig. 4.1, such that $e_i$ and $e_i'$ are independent. Notice that one of the exclusivity graphs is the complement of the other. They are also isomorphic, since the pentagon is a self-complementary graph.

Let $\{f_{ii} = e_i \wedge e_i'\}$ be the set of events defined in Corollary 4.1. The exclusivity graph of these events is the complete graph on 5 vertices, shown in Fig. 4.2.

Since $e_i$ and $e_i'$ are independent, we have

$$p\,(f_{ii}) = p(e_i)p(e_i'). \tag{4.9}$$

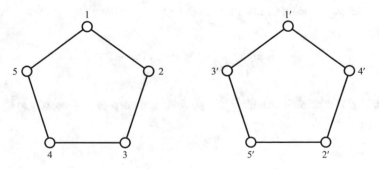

**Fig. 4.1** Exclusivity graphs of the sets of events $e_i$ and $e_i'$

**Fig. 4.2** The exclusivity
graph for the set of events
$\{f_i\}$

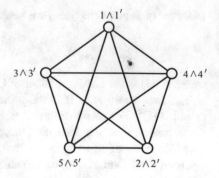

The E-principle implies

$$\sum_{i=1}^{5} p(f_{ii}) = \sum_{i=1}^{5} p(e_i)p(e_i') \leq 1. \tag{4.10}$$

Using the symmetry of the pentagon, we can assume (see Lemma 4.1 below) that
the maximum is reached when all the probabilities are equal, that is

$$p(e_i) = p(e_i') = p, \quad \forall \ i \in V(G) \tag{4.11}$$

Hence, we have

$$\sum_{i=1}^{5} p^2 = 5p^2 \leq 1 \tag{4.12}$$

which implies that

$$p \leq \frac{1}{\sqrt{5}}. \tag{4.13}$$

Now, if we substitute this value into Eq. (4.6) for $S$ we have

$$S = \sum_{i=1}^{5} p(e_i) \leq \vartheta(C_5) = \sqrt{5}. \tag{4.14}$$

Hence, if $p$ is a behaviour with $\sum_{i=1}^{5} p_i > \sqrt{5}$, then $p \otimes p \notin \mathscr{E}_E(C_5 \otimes C_5)$, and
hence $p \notin \mathscr{E}_E^2(C_5)$.                                                                   □

## 4.3   The Exclusivity Principle Forbids Behaviours Outside the Quantum Set

The idea used in the previous section to derive the quantum bound for the pentagon using the E-principle can be applied to show that there is a connection between the set of quantum behaviours for G and the set of quantum behaviours for $\overline{G}$. Yan first used it in Ref. [Yan13], where he proves the following:

**Theorem 4.3 (Yan, 2013)** *Given the set of quantum distributions for* $\overline{G}$, *the E-principle singles out the quantum maximum for* G.

*Proof* Let $\{e_i\}$ be a set of $n$ events with exclusivity graph G and $\{e'_i\}$ be a set of $n$ events with exclusivity graph $\overline{G}$, such that $e_i$ and $e'_i$ are independent. Again, if $\{f_{ii} = e_i \wedge e'_i\}$ is the set of mutually exclusive events defined in Corollary 4.1, the E-principle implies that

$$\sum_{i \in V(G)} p(f_{ii}) = \sum_{i \in V(G)} p(e_i) p(e'_i) \leq 1. \tag{4.15}$$

Suppose that the distribution $p(e'_i)$ is given by

$$p(e'_i) = |\langle \psi \mid v_i \rangle|^2. \tag{4.16}$$

Then

$$1 \geq \sum_{i \in V(G)} p(f_{ii}) = \sum_{i \in V(G)} p(e_i) p(e'_i) = \sum_{i \in V(G)} p(e_i) |\langle \psi \mid v_i \rangle|^2, \tag{4.17}$$

and hence

$$\sum_{i \in V(G)} p(e_i) \min_{j \in V(G)} \left[ |\langle \psi \mid v_j \rangle|^2 \right] \leq \sum_{i \in V(G)} p(e_i) |\langle \psi \mid v_i \rangle|^2 \leq 1, \tag{4.18}$$

which implies that

$$\sum_{i \in V(G)} p(e_i) \leq \max_{i \in V(G)} \frac{1}{|\langle \psi \mid v_i \rangle|^2}. \tag{4.19}$$

This inequality should hold for any normalised $|\psi\rangle$ and any orthogonal representation $\{|v_i\rangle\}$, and hence

$$\sum_{i \in V(G)} p(e_i) \leq \min_{|\psi\rangle, |v_i\rangle} \max_{i \in V(G)} \frac{1}{|\langle \psi \mid v_i \rangle|^2}. \tag{4.20}$$

The right-hand side is just the Lovász number of G (see [Lov79, Knu94]). Then, if $p$ is a behaviour with $\sum_{i \in V(G)} p_i > \vartheta (G)$, there is a behaviour $p' \in \mathscr{E}_Q (\overline{G})$ such that $p \otimes p' \notin \mathscr{E}_E (G \otimes \overline{G})$, and hence $p \notin \tilde{\mathscr{E}}_E (G)$. We conclude that if all quantum distributions given by Eq. (4.16) can be reached and if the E-principle holds, the maximum value of $S$ cannot exceed the quantum bound.                                    □

With the same assumptions of the previous theorem, it is possible not only to single out the quantum maximum but also the entire set of quantum behaviours.

**Theorem 4.4 (Amaral, Terra Cunha, Cabello, 2014)** *Given the quantum set $\mathscr{E}_Q (\overline{G})$, the E-principle singles out the quantum set $\mathscr{E}_Q (G)$.*

*Proof* Let $\{e_i\}$ be a set of $n$ events with exclusivity graph G and $\{e'_i\}$ be a set of $n$ events with exclusivity graph $\overline{G}$, such that $e_i$ and $e'_i$ are independent. Let $\left\{ f_{ii} = e_i \wedge e'_i \right\}$ be the set of pairwise exclusive events defined in Corollary 4.1.

Since $e_i$ and $e'_i$ are independent, we have

$$p(f_{ii}) = p_i p'_i, \tag{4.21}$$

where $p_i = p (e_i)$ and $p'_i = p (e'_i)$. The E-principle implies

$$\sum_{i \in V(G)} p_i p'_i \stackrel{\text{E}}{\leq} 1. \tag{4.22}$$

Now we use corollary 3.4 and theorem 3.5 of Ref. [GLS86]:

**Theorem 4.5** *The set* TH (G) *can be written in the following way:*

$$\text{TH} (G) = \left\{ p \in \mathbb{R}^n \,\middle|\, p_i \geq 0, \vartheta \left( \overline{G}, p \right) \leq 1 \right\}. \tag{4.23}$$

Equation (4.23) implies that, for a given $p$, inequality (4.22) will be satisfied for all $p' \in \text{TH} (\overline{G})$ if and only if $p$ belongs to TH (G). Since TH (G) $= \mathscr{E}_Q (G)$, we conclude that if the set of allowed behaviours for $\overline{G}$ is TH $(\overline{G})$ $= \mathscr{E}_Q (\overline{G})$, Theorem 4.5 implies that the behaviours for G allowed by the E-principle belong to $\mathscr{E}_Q (G)$. We conclude that if $p \notin \mathscr{E}_Q (G)$, there is a behaviour $p' \in \mathscr{E}_Q (\overline{G})$ such that $p \otimes p' \notin \mathscr{E}_E (G \otimes \overline{G})$, and hence $p \notin \tilde{\mathscr{E}}_E (G)$.                                    □

Physically, the proof above can be interpreted as follows: assuming that nature allows all quantum distributions for $\overline{G}$, the E-principle *singles out the quantum behaviours for* G.

Theorem 4.4 does not imply that the E-principle, by itself, singles out the quantum correlations for G, since we have assumed quantum theory for $\overline{G}$. Nonetheless, it is remarkable that the E-principle connects the correlations of two, a priori, completely different experiments on two completely different quantum systems. For example, if G is the $n$-cycle $C_n$ with $n$ odd, the tests of the maximum quantum violation of the corresponding noncontextuality inequalities require systems of dimension 3 [CSW10, CDLP13, LSW11, AQB$^+$13]. However, the tests of the maximum quantum violation of the noncontextuality inequalities with exclusivity graph $\overline{C_n}$ require systems of dimension that grows with $n$ [CDLP13]. Similarly, while two qubits are enough for a test of the maximum quantum violation of the CHSH inequality, the complementary test is a noncontextuality inequality (not a Bell inequality) that requires a system of, at least, dimension 5 [Cab13a].

An important consequence of Proposition 4.4 is that the larger the quantum set of G, the smaller the quantum set for $\overline{G}$, since each probability allowed for G becomes a restriction on the possible probabilities for $\overline{G}$. Such duality gets stronger when G is a self-complementary graph.

**Theorem 4.6 (Amaral, Terra Cunha, Cabello, 2014)** *If G is a self-complementary graph, the E-principle, by itself, excludes any set of probability behaviours strictly larger than the quantum set.*

*Proof* Let X be a set of behaviours containing $\mathscr{E}_Q$ (G) and let $p \in X \setminus \mathscr{E}_Q$ (G). By Proposition 1, there is at least one $p' \in \mathscr{E}_Q (\overline{G})$ such that

$$\sum_{i \in V(G)} p_i p_i' > 1. \tag{4.24}$$

Since G is self-complementary, after a permutation on the entries given by the isomorphism between G and $\overline{G}$, $p'$ becomes an element of $\mathscr{E}_Q$ (G) and hence we can assume that $p$ and $p'$ belong to X. Expression (4.24) implies that this set is in contradiction with the E-principle. □

The fact that the E-principle is sufficient for pinning down the quantum behaviours as the maximal set of behaviours for any self-complementary graph, given that the entire quantum set is possible, means that the E-principle is able to single out the quantum behaviours for a large number of nonequivalent exclusivity scenarios, including the KCBS scenario.

The hypothesis in Theorem 4.3 can be weakened for vertex transitive graphs. Instead of assuming the entire set of quantum behaviours for $\overline{G}$, the same result can be proven given only the quantum maximum for $\overline{G}$. The exclusivity graphs of many interesting inequalities including CHSH [CHSH69], KCBS [KCBS08], the $n$-cycle inequalities [CSW10, CDLP13, LSW11, AQB$^+$13], and the antihole inequalities [CDLP13] are vertex transitive.

**Definition 4.4** A graph is *vertex transitive* if for any pair $i, j \in V$ (G) there is $\phi \in$ Aut (G) such that $i = \phi(j)$, where Aut (G) is the group of automorphisms of G (*i.e.*, the permutations $\psi$ of the set of vertices such that $i, j \in V$ (G) are adjacent if and only if $\psi(i), \psi(j)$ are adjacent).

**Theorem 4.7 (Amaral, Terra Cunha, Cabello, 2014)** *If G is a vertex-transitive graphs, given the quantum maximum for $\overline{G}$, the E-principle singles out the quantum maximum for G.*

A sequence of three lemmas proves the result. First we prove that the quantum maximum for $S$ is assumed at a symmetric configuration. Then we prove that the product of the quantum maxima for $G$ and $\overline{G}$ is bounded from above by the number of vertices of $G$, and the same from below.

**Lemma 4.1** *If G is a vertex-transitive graph, the quantum maximum for*

$$S = \sum_{i \in V(G)} p_i \tag{4.25}$$

*is attained at the constant behaviour $p_i = p_{max}$.*

*Proof* Let $p = (p(e_1), p(e_2), \ldots, p(e_n))$ be a behaviour reaching the maximum. Given an automorphism $\phi \in \mathrm{Aut}\,(G)$, consider the behaviour $p_\phi$ defined as $p_\phi(e_i) = p(e_{\phi(i)})$. This is a valid quantum behaviour, also reaching the maximum for $S$. Define the behaviour

$$q = \frac{1}{A} \sum_{\phi \in \mathrm{Aut}(G)} p_\phi, \tag{4.26}$$

where $A = \#\mathrm{Aut}\,(G)$. This behaviour also reaches the maximum for $S$. Since G is vertex transitive, given any two vertices of G, $i$ and $j$, there is an automorphism $\psi$ such that $\psi(i) = j$. Then,

$$q(e_j) = q(e_{\psi(i)})$$

$$= \frac{1}{A} \sum_{\phi \in \mathrm{Aut}(G)} p_\phi(e_{\psi(i)})$$

$$= \frac{1}{A} \sum_{\phi \in \mathrm{Aut}(G)} p\left(e_{\phi \circ \psi(i)}\right)$$

$$= \frac{1}{A} \sum_{\phi' \in \mathrm{Aut}(G)} p_{\phi'}(e_i)$$

$$= q(e_i). \tag{4.27}$$

$\square$

**Lemma 4.2** *If G is a vertex-transitive graph on n vertices, then the E-principle implies that the quantum maxima of S for G and $\overline{G}$ obey*

$$\beta_Q(G)\,\beta_Q\big(\overline{G}\big) \overset{E}{\leq} n. \tag{4.28}$$

*Proof* Lemma 4.1 applies for both, G and $\overline{G}$, giving $np_{max} = \beta_Q(G)$ and $np'_{max} = \beta_Q(\overline{G})$. Inequality (4.22) for these extremal behaviours reads

$$\sum_i p_i\, p'_i = n\, p_{max}\, p'_{max} \overset{E}{\leq} 1, \qquad (4.29)$$

which proves the result.                                                                                      □

**Lemma 4.3** *If* G *is a vertex-transitive graph on* n *vertices, then*

$$\beta_Q(G)\, \beta_Q(\overline{G}) \geq n. \qquad (4.30)$$

*Proof* When we recall that the graph approach identifies the quantum maximum with the Lovász number, as proven in Theorem 3.6, we have that

$$\vartheta\,(G)\, \vartheta\,(\overline{G}),$$

$$\vartheta(\overline{G}) = \beta_Q\,(\overline{G}), \qquad (4.31)$$

and since for vertex-transitive graphs $\vartheta\,(G)\, \vartheta\,(\overline{G}) \geq n$ (lemma 23 in Ref. [Knu94]), the lemma follows.                                                                       □

Proposition 4.7 opens the door to experimentally discard non-quantum behaviours. Specifically, Lemma 4.2 implies that we can test if the maximum value of behaviours with exclusivity graph G goes beyond the quantum maximum without violating the E-principle by performing an independent experiment testing behaviours with exclusivity graph $\overline{G}$ and experimentally reaching its quantum maximum [Cab13a]. A violation of the quantum bound for $\overline{G}$ in any laboratory would imply the impossibility of reaching the quantum maximum for G in any other laboratory. From the fundamental point of view, it resembles some discussions on *which-way interferometers* and *quantum erasers* [WCaPM03]: it does not matter if someone actually does an experiment saturating $\overline{G}$ inequalities; the mere possibility of doing so, assured by quantum theory, is sufficient for imposing the bound of G inequality.


## 4.4   Other Graph Operations

We have seen in the previous section that using the operation of taking the OR-product with the complement of a graph and the E-principle, we are able to explain the quantum bound and the quantum set for many different noncontextuality inequalities. In this section, we study if something similar is possible using other graph operations.

### 4.4.1  Direct Cosum

**Definition 4.5** Given two graphs $G'$ and $G''$ we define the *direct cosum* $G$ of $G'$ and $G''$ as the graph with $V(G) = V(G') \sqcup V(G'')$ and such that $\{i, j\} \in E(G)$ iff $\{i, j\} \in E(G')$, or $\{i, j\} \in E(G'')$, or $i \in V(G')$ and $j \in V(G'')$.

The direct cosum of $G'$ and $G''$ consists of the graph obtained from joining all vertices of $G'$ with all vertices of $G''$. This operation applied to two copies of $C_5$ is illustrated[3] in Fig. 4.3. The result below is a well-known fact and can be found

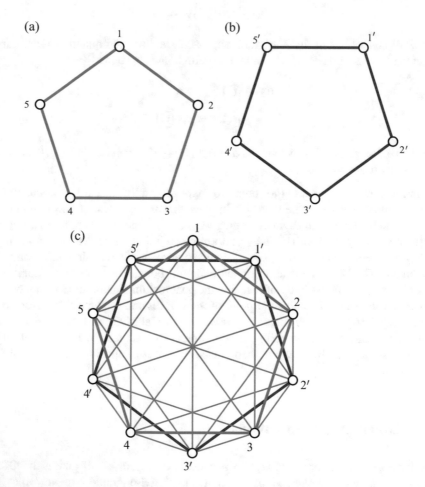

**Fig. 4.3** Two copies of the pentagon C5 (**a, b**) and their direct cosum (**c**), which can be recognised as the circulant graph $C_{10}(1, 2, 3, 5)$

---

[3]For $C_5$, this operation is equivalent to applying the duplication defined is Sect. 4.4.2 and complementation, but this is not true in general. For general graphs $G'$ and $G''$, $G = \overline{\overline{G'} + \overline{G''}}$,

on Ref. [Knu94], but we repeat it here to reinforce the connections with quantum theory.

**Lemma 4.4** *If* G *is the direct cosum of* G' *and* G'',

$$\vartheta\left(G\right) = \max\left\{\vartheta\left(G'\right), \vartheta\left(G''\right)\right\}. \tag{4.32}$$

*Proof* Let $\{|v_i\rangle\}$ be an orthonormal representation for G and $|\psi\rangle$ be a unit vector in the same vector space. Every vertex of G' is exclusive to all vertices of G'', which means that the vectors of the orthonormal representation for G generate a subspace V' orthogonal to the subspace V'' generated by the vectors of the orthonormal representation for G''. Hence, we can decompose $|\psi\rangle$ as a sum of two orthonormal vectors:

$$|\psi\rangle = a\left|\psi'\right\rangle + b\left|\psi''\right\rangle, \quad \left|\psi'\right\rangle \in V', \quad \left|\psi''\right\rangle \in V'', \quad |a|^2 + |b|^2 = 1. \tag{4.33}$$

With these definitions we have

$$\sum_{i \in V(G)} |\langle \psi \mid v_i \rangle|^2 = |a|^2 \left(\sum_{i \in V(G')} |\langle \psi' \mid v_i \rangle|^2\right) + |b|^2 \left(\sum_{i \in V(G'')} |\langle \psi'' \mid v_i \rangle|^2\right) \tag{4.34}$$

and then

$$\vartheta\left(G\right) \leq \max\left\{\vartheta\left(G'\right), \vartheta\left(G''\right)\right\}. \tag{4.35}$$

Suppose $\max\left\{\vartheta\left(G'\right), \vartheta\left(G''\right)\right\} = \vartheta\left(G'\right)$. Let $\{|v_i'\rangle\}$ be a Lovász optimal representation for G' and $|\psi\rangle$ the unit vector achieving $\vartheta\left(G'\right)$. Let $\{|v_i''\rangle\}$ be any Lovász representation for G''. The set of vectors $\{|v_i'\rangle \oplus 0, 0 \oplus |v_i''\rangle\}$ is an optimal Lovász representation for G and the unit vector $|\psi\rangle \oplus 0$ achieves the upper bound in Eq. (4.35). □

As a consequence of the above result, we have:

**Corollary 4.2** *If the E-principle rules out violations above the quantum maximum for* G, *it also rules out violations above the quantum maximum for its direct cosum with any other graph* G' *such that* $\vartheta\left(G'\right) \leq \vartheta\left(G\right)$. *In particular, it rules out violations above the quantum maximum for the direct cosum of* G *with itself.*

In fact, as we have seen in the proof of Lemma 4.4, if $\vartheta\left(G'\right) \leq \vartheta\left(G\right)$, the Lovász optimal representation for the direct cosum of G' and G can be obtained from a Lovász optimal representation of G, and if the E-principle constrains the latter, it will also constrain the former.

---

where the direct sum + of graphs is defined by the disjoint union of vertices and edges. Actually, the expression above justifies the name direct cosum.

### 4.4.2  Twinning, Partial Twinning and Duplication

We can also consider graphs obtained from two copies of G by adding some of
the edges between the vertices of each copy of G but not all of them. One of this
graphs is the graph T (G) obtained if we consider two copies of G with the same
labeling and join the vertices of one of the copies with the exclusive vertices of
the other copy. Figure 4.4a shows this operation applied to the pentagon. We call
this operation *twinning*, since the graph associated to T (G) is the one obtained by
duplicating all the vertices of G.

**Definition 4.6** Consider two copies of a graph G, where the vertices of the first
copy are denoted by $\{i\}$ and the vertices of the second copy are denoted by $\{j'\}$. The
*twinning* of G is the graph T (G) such that $V[T(G)] = V(G) \sqcup V(G) = \{i\} \cup \{j'\}$
and such that $E[T(G)] = E \cup E' \cup E''$ where

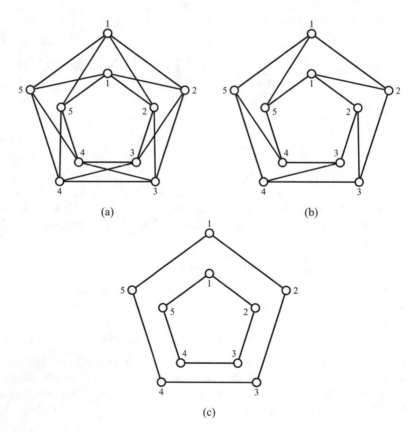

(a)                                                          (b)

(c)

**Fig. 4.4** (a) The twinning of $C_5$. (b) A partial twinning of $C_5$. (c) The direct sum of two copies of
$C_5$

$$E = \{\{i, j\} \,|\, \{i, j\} \in E\,(G)\}$$

$$E' = \{\{i', j'\} \,|\, \{i, j\} \in E\,(G)\}\,.$$

$$E'' = \{\{i, j'\} \,|\, \{i, j\} \in E\,(G)\}\,. \tag{4.36}$$

The graphs obtained by replacing $E''$ by one of its proper subsets are called *partial twinnings* of G.

Figure 4.4a shows the twinning of $C_5$. Partial twinnings of $C_5$ can be obtained by removing any of the ten edges present in Fig. 4.4a and absent in Fig. 4.4c. Figure 4.4b is just a particular case of this.

**Theorem 4.8** *The Lovász number of* $T\,(G)$ *is given by* $\vartheta\,[T\,(G)] = 2\vartheta\,(G)$. *Moreover, if* $G'$ *is any partial twinning of* G, *then* $\vartheta\left(G'\right) = 2\vartheta\,(G)$.

*Proof* Let $G'$ be the twinning or any partial twinning of G. The upper bound $\vartheta\left(G'\right) \leq 2\vartheta\,(G)$ comes from the fact that each copy of G is an induced subgraph of $G'$ and this implies that every orthonormal representation for $G'$ gives an orthonormal representation for each copy of G. On the other hand, given an optimal orthonormal representation $\{|v_i\rangle\}_{i=1}^{n}$ for G with handle $|\psi\rangle$ achieving $\vartheta\,(G)$, the set of vectors $\{|v_i\rangle\}_{i=1}^{2n}$ with $|v_i\rangle = |v_{i+n}\rangle$ form an orthonormal representation for $G'$ such that, with handle $|\psi\rangle$, gives

$$\sum_{i=1}^{2n} |\langle v_i \,|\, \psi\rangle|^2 = 2\vartheta\,(G)\,. \tag{4.37}$$

This implies that $\vartheta\left(G'\right) \geq 2\vartheta\,(G)$. □

From Theorem 4.8, we have:

**Corollary 4.3** *If the E-principle singles out the quantum maximum for a graph* G, *it also singles out the quantum maximum for its twinning and all its partial twinnings.*

The extreme case of partial twinning, presented for the pentagon in Fig. 4.4c, is also called the direct sum of G with itself [Knu94]. We call this operation *duplication*[4] of G. We can apply this same operation on two different graphs $G'$ and $G''$, obtaining the direct sum $G = G' + G'$, which is the graph with vertex-set $V\left(G'\right) \sqcup V\left(G''\right)$ and such that $\{i, j\} \in E\,(G)$ if either $\{i, j\} \in E\left(G'\right)$ or $\{i, j\} \in E\left(G''\right)$. Clearly $\vartheta\,(G) = \vartheta\left(G'\right) + \vartheta\left(G''\right)$, and we also have the trivial result that if the E-principle singles out the quantum maximum for $G'$ and $G''$ it also singles out the quantum maximum for G.

---

[4]Although the term *duplication* is sometimes used to refer to a different graph operation than the one we define here, we choose this term because its physical interpretation: for exclusivity graph, the duplication, as defined above, represents two realisations of the same experiment.

### 4.4.3   Vertex-Transitive Graphs Obtained from $C_5$

Applying the operations above to $C_5$, for which the E-principle singles out the quantum maximum, we can explain the quantum maximum for almost all vertex-transitive graphs with 10 vertices.

Among the vertex-transitive graphs on 10 vertices, only eight have $\vartheta\,(G) > \alpha\,(G)$, the circulant graphs $Ci_{10}(1, 2, 3, 5)$, $Ci_{10}(1, 4)$, $Ci_{10}(2, 5)$, $Ci_{10}(2, 3, 5)$, $Ci_{10}(1, 2, 3)$, $Ci_{10}(1, 2)$, $Ci_{10}(1, 2, 5)$ and the Johnson graph $J(5, 2)$ [Wika, Wikb].

**Theorem 4.9 (Amaral (2014))** *The quantum maximum for the graphs $J(5, 2)$, $Ci_{10}(1, 2, 3, 5)$, $Ci_{10}(1, 4)$, $Ci_{10}(2, 5)$, $Ci_{10}(2, 3, 5)$ and $Ci_{10}(1, 2, 3)$ is the maximum value allowed by the E-principle.*[5]

*Proof* Since $\vartheta\,(J(5, 2)) = \alpha^*\,(J(5, 2))$, the E-principle by itself explains the quantum maximum for this graph. The graph $Ci_{10}(1, 2, 3, 5)$ is the direct cosum of $C_5$ with itself, $Ci_{10}(1, 4)$ is the twinning of $C_5$, $Ci_{10}(2, 5)$ is a partial twinning of $C_5$, $Ci_{10}(2, 3, 5)$ is isomorphic to the complement of $Ci_{10}(1, 4)$, and $Ci_{10}(1, 2, 3)$ is the complement of $Ci_{10}(2, 5)$. Hence, the result follows from Theorem 4.7 and Corollaries 4.2 and 4.3.                                                                    □

## 4.5   The Exclusivity Principle in the Exclusivity-Hypergraph Approach

Also in the Exclusivity-Hypergraph Approach, the E-principle imposes extra conditions on the behaviours $p$ of a given contextuality scenario H, as we have seen in Sect. 3.4.4. Once more, we can combine different behaviours in order to explore the activation aspects of the E-principle. For this purpose, we first define an adequate notion of *product* of two contextuality scenarios in the exclusivity-hypergraph approach.

### 4.5.1   The Foulis-Randall Product

**Definition 4.7** The *Foulis-Randall (FR) product* of two contextuality scenarios $H_A$ and $H_B$ is the contextuality scenario $H_{AB} := H_A \otimes H_B$ with vertex-set given by

$$V\,(H_{AB}) = V\,(H_A) \times V\,(H_B) \tag{4.38}$$

---

[5]For more details, see Ref. [Ama14]. This work has been done in collaboration with Adán Cabello, during a visit to Seville in 2013.

**Fig. 4.5** The exclusivity
hypergraph $H_{1,2,2}$

and edge-set given by

$$E\left(H_{AB}\right) = E_{A \to B} \cup E_{B \to A}, \tag{4.39}$$

where

$$E_{A \to B} = \left\{ \cup_{a \in e_A} \{a\} \times f(a) \,\middle|\, e_A \in E\left(H_A\right),\ f : e_A \to E\left(H_B\right) \right\} \tag{4.40}$$

and

$$E_{B \to A} = \left\{ \cup_{b \in e_B} f(b) \times \{b\} \,\middle|\, e_B \in E\left(H_B\right),\ f : e_B \to E\left(H_A\right) \right\}. \tag{4.41}$$

Intuitively, a hyperedge $e \in E_{A \to B}$ corresponds to the following situation: first, a measurement $e_A \in E\left(H_A\right)$ is chosen for the first party. Then, a measurement for the second party is chosen as a function $f$ of the first party's measurement choice and outcome. This defines a joint measurement that can be interpreted as the first party measuring first and then communicating her outcome to the second party, who will choose the measurement to be conducted as a function of the information communicated by the first party. The same interpretation holds for the hyperedges $e \in E_{B \to A}$ with parties $A$ and $B$ exchanged. The hyperedges that belong to $E_{A \to B} \cap E_{B \to A}$ are of the form $e = e_A \times e_B$ and can be interpreted as parties $A$ and $B$ measuring without any communication between them.

Given behaviours $p_A \in \mathcal{G}\left(H_A\right)$ and $p_B \in \mathcal{G}\left(H_B\right)$, $p_A \otimes p_B$ is a behaviour in $\mathcal{G}\left(H_{AB}\right)$ that represents an independent realisation of $p_A$ and $p_B$. In general there will be behaviours in $\mathcal{G}\left(H_{AB}\right)$ that are not of this form, as expected, since there may be some correlations between the parties.

**Theorem 4.10** *The exclusivity graph of the FR-product $H_{AB}$ is the OR product of the exclusivity graphs for $H_A$ and $H_B$.*

The proof of this result can be found in Ref. [AFLS15].

*Example 4.1 (Bell Scenario)* The Bell scenario $(n, k, m)$ consists of $n$ parties having each $k$ measurements with $m$ outcomes per measurement. The hypergraph associated to this scenario is

$$H_{n,k,m} = H_{1,k,m}^{\otimes n}, \tag{4.42}$$

the FR-product of $n$ copies of the scenario introduced in Example 3.3 (Fig. 4.5).

The set $\mathcal{G}\left(H_{n,k,m}\right)$ consists of the set of all nosignalling behaviours in the Bell scenario $(n, k, m)$. More generally, we have:

**Fig. 4.6** The exclusivity
hypergraph
$H_{2,2,2} = H_{1,2,2} \otimes H_{1,2,2}$. The
hyperedges in which first
party measures first are
depicted in red; the
hyperedges in which second
party measures first are
depicted in green; the
hyperedges representing
simultaneous measurements
are depicted in gray

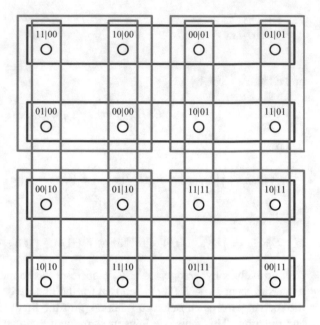

**Theorem 4.11** *The set $\mathscr{G}(H_{AB})$ is the set of nosignalling models between parties A and B.*

See Ref. [AFLS15] for a proof of this result.

*Example 4.2 (The CHSH Scenario)* Let us see in detail what happens for the CHSH scenario (2, 2, 2) introduced in Sect. 2.3.1. In this case the scenario is defined by the hypergraph

$$H_{2,2,2} = H_{1,2,2} \otimes H_{1,2,2}. \tag{4.43}$$

The edges representing simultaneous measurements are of the form $e_A \times e_B$, where parties $A$ and $B$ will choose a measurement to perform independently of the other party outputs. We have a total of four of such edges:

$$
\begin{aligned}
&\{00|00, 01|00, 10|00, 11|00\}, \\
&\{00|01, 01|01, 10|01, 11|01\}, \\
&\{00|10, 01|10, 10|10, 11|10\}, \\
&\{00|11, 01|11, 10|11, 11|11\}.
\end{aligned}
\tag{4.44}
$$

The corresponding hyperegdes are shown in gray in Fig. 4.6. If the first party measures first and the second party's measurement depends on the output of first party's measurement, the events in the corresponding edge will be of the form $ab|xf(a)$. The function $f$ cannot be constant, otherwise we would recover one of the edges already listed in (4.44). Hence, we must have $f(a) = a$ or $f(a) = 1 - a$.

In the first case we have the edges

$$\{00|00, 01|00, 10|01, 11|01\},$$
$$\{00|10, 01|10, 10|11, 11|11\}, \tag{4.45}$$

and in the second case we have the edges

$$\{00|01, 01|01, 10|00, 11|00\},$$
$$\{00|11, 01|11, 10|10, 11|10\}. \tag{4.46}$$

The corresponding hyperedges are shown in red in Fig. 4.6. Exchanging the roles of the parties we get the edges where the second party measures first and the first party's measurement depends on the output of the second party's measurement. The events in the corresponding edge will be of the form $ab|f(b)y$, where $f(b) = b$ or $f(b) = 1 - b$. This will lead to the edges

$$\{00|00, 10|00, 01|10, 11|10\},$$
$$\{00|01, 10|01, 01|11, 11|11\},$$
$$\{00|10, 10|10, 01|00, 11|00\}, \tag{4.47}$$
$$\{00|11, 10|11, 01|01, 11|01\}.$$

The corresponding hyperedges are shown in green in Fig. 4.6.

It is important to notice that the FR-product of contextuality scenarios is not associative, that is, there are scenarios $H_A, H_B$, and $H_C$ such that $(H_A \otimes H_B) \otimes H_C \neq H_A \otimes (H_B \otimes H_C)$. In Ref. [AFLS15], the authors define the notions of *minimal* and *maximal* Foulis-Randall product, in order to take those differences into account. Nonetheless, these differences are not observable on the operational level, in the sense that the classical, quantum and E-principle sets are identical irrespective of the product considered. Hence, given a family of contextuality scenarios $H_i$, we will denote by $\otimes_i H_i$ any of the products of $H_i$, as defined in Ref. [AFLS15].

## 4.5.2   Activation Effects of the E-Principle

Just as in the exclusivity graph approach, we assume that arbitrary combinations of behaviours satisfying the E-principle should also satisfy this principle in the large combined scenario. This, once more, leads us to the following refinements of the E-principle:

**Definition 4.8** We denote by $\mathcal{H}_E^n(H)$ the set of behaviours $p \in \mathcal{G}(H)$ such that

$$p^{\otimes n} \in \mathcal{H}_E(H^{\otimes n}). \tag{4.48}$$

Furthermore, we define

$$\mathcal{H}_E^\infty (H) := \bigcap_{n \in \mathbb{N}} \mathcal{H}_E^n (H) . \tag{4.49}$$

These sets can be characterised in terms of graph invariants.

**Definition 4.9** The *strong product* of graphs $G_1$ and $G_2$ is the graph $G_1 \boxtimes G_2$ whose vertex-set is $V(G_1 \boxtimes G_2) = V(G_1) \times V(G_2)$ and such that $(i_1, i_2)$ and $(j_1, j_2)$ belong to $E(G_1 \boxtimes G_2)$ if one of the following holds:

1. $i_1 = j_1$ and $(i_2, j_2) \in E(G_2)$;
2. $i_2 = j_2$ and $(i_1, j_1) \in E(G_1)$;
3. $(i_1, j_1) \in E(G_1)$ and $(i_2, j_2) \in E(G_2)$.

**Definition 4.10** The *Shannon capacity*[6] of a graph $G$ is defined from the independence number of its strong products with itself:

$$\Theta(G) = \sup_n \sqrt[n]{\alpha\left(G^{\boxtimes n}\right)} = \lim_{n \to \infty} \sqrt[n]{\alpha\left(G^{\boxtimes n}\right)}, \tag{4.50}$$

where $G^{\boxtimes n}$ is the strong product of $n$ copies of $\overline{G}$.

**Theorem 4.12** *For a behaviour $p \in \mathcal{G}(H)$,*

1. $p \in \mathcal{H}_E^n (H)$ *if, and only if,*

$$\alpha\left(\overline{G}^{\boxtimes n}, p^{\otimes n}\right) = 1. \tag{4.51}$$

2. $p \in \mathcal{H}_E^\infty (H)$ *if, and only if,*

$$\Theta\left(\overline{G}, p\right) = 1. \tag{4.52}$$

Item 1 is a consequence of Theorem 3.23 and the fact that $\overline{G^{\otimes n}} = \overline{G}^{\boxtimes n}$. Item 2 follows from the definition of $\Theta$.

**Theorem 4.13** *For every contextuality scenario H,*

$$\mathcal{H}_{AFLS}^1 (H) \subset \mathcal{H}_E^\infty (H) . \tag{4.53}$$

*Moreover, there are contextuality scenarios for which*

$$\mathcal{H}_{AFLS}^1 (H) \subsetneq \mathcal{H}_E^\infty (H) . \tag{4.54}$$

---

[6]The computational complexity of the Shannon capacity is unknown, and even its value for certain small graphs such as $C_7$ remains unknown. The Lováz number was developed precisely to provide an upper bound to $\Theta$ that can be computed efficiently [Lov79].

*Proof* If $p \in \mathcal{H}^1_{AFLS}(H)$, then $\vartheta(\overline{G}, p) = 1$. Since

$$\Theta(\overline{G}, p) \leq \vartheta(\overline{G}, p), \tag{4.55}$$

we have that $\Theta(\overline{G}, p) \leq 1$. Normalisation of $p$ implies that $\Theta(\overline{G}, p) = 1$, which in turns implies that $p \in \mathcal{H}^\infty_E(H)$. An example in Ref. [AFLS15] proves the second part of the theorem. □

In Ref. [AFLS15] the authors explore the activation effects of the E-principle. In fact, they show that there are contextuality scenarios $H_A$ and $H_B$ for which

$$\mathcal{H}^\infty_E(H_A) \otimes \mathcal{H}^\infty_E(H_B) \nsubseteq \mathcal{H}^1_E(H_{AB}). \tag{4.56}$$

They also show that for some scenarios an even more drastic result holds: $\mathcal{H}^\infty_E(H)$ may not even be a convex set.

We have, so far, considered the consequences of the E-principle over several copies of the behaviour $p$. We can also consider combinations of $p$ with other arbitrary behaviours $q$.

**Definition 4.11 (Complete Exclusiveness)** We denote by $\tilde{\mathcal{H}}_E(H)$ the set of behaviours $p \in \mathcal{G}(H)$ such that

$$p \otimes q \subset \mathcal{H}_E(H \otimes H') \tag{4.57}$$

for every behaviour $q \in \mathcal{H}_Q(H')$. and every contextuality scenario $H'$.

**Theorem 4.14** *For every contextuality scenario and for every $n \in \mathbb{N}$ we have that*

$$\tilde{\mathcal{H}}_E(H) = \mathcal{H}^1_{AFSL}(H). \tag{4.58}$$

For a proof of this result, see Ref. [AFLS15]. Notice that this result implies that in the exclusivity-hypergraph approach the E-principle *does not* characterise the quantum set, since $\mathcal{H}_Q(H) \subsetneq \mathcal{H}^1_{AFLS}(H)$. This shows that requiring normalisation of $p$ implies extra constraints over the set of quantum behaviours that cannot be captured by the E-principle alone. Nonetheless, this result restricts considerably the set of allowed behaviours, since $\mathcal{H}^1_{AFLS}(H)$ is often close to $\mathcal{H}_Q(H)$. Indeed, one can prove the following result:

**Corollary 4.4** *The maximum allowed by the E-principle for the Collins-Gisin-Linden-Massar-Popescu (CGLMP) inequalities [CGL$^+$02] in the $(2, 2, d)$ Bell scenario is equal to the quantum maximum.*

*Proof* In Ref. [NPA08] the authors show that the maximum for the CGLMP inequalities attained for behaviours in $\mathcal{H}^1_{AFLS}(H)$ is equal to the quantum maximum. The result follows from Theorem 4.14. □

## 4.6   Final Remarks

In this chapter, we have shown that the E-principle is able to single out the quantum maximum and even the entire set of quantum distributions in many different situations. The results found so far are listed below.

1. The Exclusivity principle directly explains the quantum maximum for all graphs with $\vartheta$ (G) $= \alpha^*$ (G) [CSW10];
2. Given the set of quantum distributions for $\overline{G}$, the E-principle explains the entire set of quantum behaviours for G, as shown in Theorem 4.4 [ATC14];
3. The E-principle, applied to two copies of the graph, explains the entire set of quantum behaviours for self-complementary graphs, including the pentagon, the simplest graph exhibiting quantum contextuality, as shown in Theorem 4.6 [Cab13b, ATC14];
4. Given the quantum maximum for $\overline{G}$, the Exclusivity principle explains the quantum maximum for any vertex-transitive graph G, as shown in Theorem 4.7 [ATC14];
5. The E-principle explains the quantum maximum for at least six of the eight vertex-transitive graphs with 10 vertices, as shown in Theorem 4.9;
6. Either applied to two copies of the exclusivity graph of the CHSH inequality [FSA+13] or of a simpler inequality [Cab13b], the E-principle excludes PR boxes and provides an upper bound to the maximum violation of the CHSH inequality which is close to the Tsirelson bound;
7. The E-principle rules out all extremal non-quantum distributions in the $(2, 2, d)$ Bell scenarios [FSA+13];
8. When applied to the OR product of an infinite number of copies, there is strong numerical evidence that the E-principle singles out the maximum quantum violation of the noncontextuality inequalities whose exclusivity graph is the complement of odd cycles on $n \geq 7$ vertices [CDLP13]. Indeed, it might be also the case that, when applied to an infinite number of copies, the E-principle singles out the Tsirelson bound of the CHSH inequality [FSA+13, Cab13b].
9. The E-principle *does not* characterise the quantum set in the exclusivity-hypergraph scenario. This should not be a surprise, since the E-principle is related only to exclusivities and has no relation with other constraints such as compatability structures and normalisation. Applied to general products of two scenarios, the E-principle singles out $\mathscr{H}^1_{AFLS}$ (H), which is close to the quantum set in many situations. For example, this results implies that the maxima for the CGLMP inequalities allowed by the E-principle are equal to the quantum maxima.

The simplest vertex-transitive graphs are shown in Fig. 4.7. The strength of the Exclusivity principle can be very well exemplified if we analyse what it predicts for those graphs. For G $= C_5$, the E-principle explains the entire set of quantum behaviours. For $C_7$ and $C_9$, there are strong numerical evidences that it explains the quantum maximum.[7] If this is indeed the case, we can also explain the quantum

---

[7] A. Cabello, private communication.

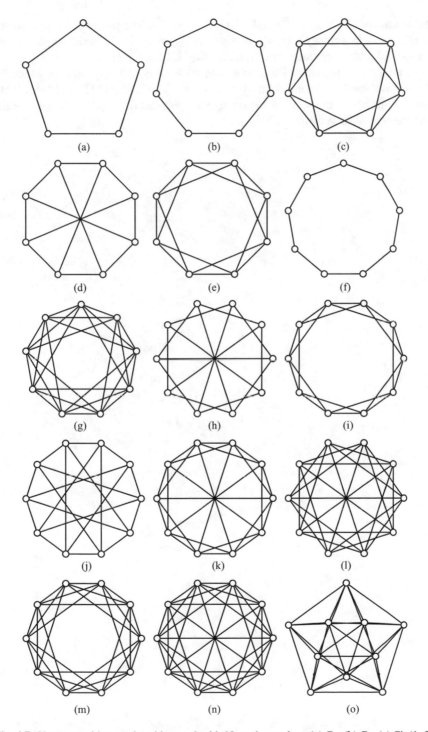

**Fig. 4.7** Vertex transitive graphs with $\alpha < \vartheta$ with 10 vertices or less. (**a**) $C_5$. (**b**) $C_7$. (**c**) $Ci_7(1, 2)$. (**d**) $Ci_8(1, 4)$. (**e**) $Ci_8(1, 2)$. (**f**) $C_9$. (**g**) $Ci_9(1, 2, 3)$. (**h**) $Ci_{10}(2, 5)$. (**i**) $Ci_{10}(1, 2)$. (**j**) $Ci_{10}(1, 4)$. (**k**) $Ci_{10}(1, 2, 5)$. (**l**) $Ci_{10}(2, 3, 5)$. (**m**) $Ci_{10}(1, 2, 3)$. (**n**) $Ci_{10}(1, 2, 3, 5)$. (**o**) $J(5, 2)$

maximum for $Ci_7(1, 2) = \overline{C_7}$ and $Ci_9(1, 2, 3) = \overline{C_9}$. The E-principle explains the quantum maximum for $Ci_8(1, 4)$, the exclusivity graph of the CHSH inequality, and it also explains the quantum maximum for $Ci_8(1, 2) = \overline{Ci_8(1, 4)}$.

Among the vertex-transitive graphs with 10 vertices, the only ones for which we can still not explain the quantum maximum are $Ci_{10}(1, 2)$ and $Ci_{10}(1, 2, 5)$. If we can do that for one of them, the E-principle will explain the other, since they are the complement of each other.

# Appendix A
# State-Independent Proofs
# of the Bell-Kochen-Specker Theorem

In this chapter we present a brief review of the state-independent proofs of the Bell-Kochen-Specker Theorem.

One of the first to challenge the idea of completing quantum theory was von Neumann, and his argument appears in his pioneer rigorous mathematical formulation of quantum theory [vN55]. Although one of his assumptions was too strong, his result was a landmark in foundations of physics.

We then present Gleason's lemma. Although his main result is motivated by a problem in foundations of quantum theory, Gleason's goal had in principle nothing to do with completions of quantum theory, which are not even mentioned in his paper. Nevertheless, one of his lemmas provides an elegant proof of the Bell-Kochen-Specker theorem. This fact was noticed only after Kochen and Specker had proven this result by other means.

The advantage of Kochen and Specker's proof is that, contrary to Gleason's argument, it requires only a finite set of one-dimensional projectors. The Kochen-Specker original proof is presented, along with other simple additive and multiplicative proofs. We finish this chapter presenting a contextual completion of a quantum probabilistic model, showing that the noncontextualitiy hypothesis is indeed crucial to discard the possibility of completions.

## A.1 von Neumann

von Neumann was the first to rigorously establish a mathematical formulation for quantum theory, published in his 1932 work *Mathematische Grundlagen der Quantenmechanik*, and later translated to English in 1955 [vN55]. His rigorous approach allowed him to challenge the ideas of completion of quantum theory.

© The Author(s), under exclusive licence to Springer Nature Switzerland AG 2018    101
B. Amaral, M. Terra Cunha, *On Graph Approaches to Contextuality and their Role in Quantum Theory*, SpringerBriefs in Mathematics,
https://doi.org/10.1007/978-3-319-93827-1

He derived the quantum formula

$$\langle O \rangle = \text{Tr}\,(\rho O) \tag{A.1}$$

for the expectation value of a measurement from a few general assumptions about the expectation-value function. Then, we can prove that there is no deterministic state consistent with this formula, and hence that completions of quantum theory are impossible. Although one of his assumptions was too strong, as we explain later, his result was breakthrough that preceded a series of elegant proofs of the Bell-Kochen-Specker theorem.

### A.1.1   von Neumann's Assumptions

Given a specific type of system in a probability theory, every state defines an *expectation-value function*:

$$\langle \, \rangle : \mathcal{M} \longrightarrow \mathbb{R} \tag{A.2}$$

where $\mathcal{M}$ stands for the set of measurements in the completion. Instead of focusing on the possible states, von Neumann was interested in the properties of these functions, and stated a number of requirements he believed to be natural impositions on them.

**Definition A.1**  An expectation-value function $\langle \, \rangle : \mathcal{M} \longrightarrow \mathbb{R}$ is *deterministic* if

$$\langle M^2 \rangle = \langle M \rangle^2 \tag{A.3}$$

for every measurement M allowed in the model.

Deterministic functions are the ones that come from deterministic states, that is, those for which one of the outcomes has probability one for every measurement $M \in \mathcal{M}$.

**Definition A.2**  An expectation-value function $\langle \, \rangle : \mathcal{M} \longrightarrow \mathbb{R}$ is called *pure* if

$$\langle \, \rangle = p\langle \, \rangle' + (1 - p)\langle \, \rangle'', \ 0 < p < 1, \tag{A.4}$$

implies $\langle \, \rangle = \langle \, \rangle' = \langle \, \rangle''$.

Pure functions are the ones that cannot be written as a convex sum of others. A function $\langle \, \rangle$ is pure if, and only if, the state that defines it is a pure state of the system. Every deterministic function is pure, but the converse is not always true. For example, in quantum theory, pure functions are the ones defined by one-dimensional projectors, while there is no deterministic function. In a noncontextual completion of a probabilistic model the two notions coincide.

In quantum theory, every measurement M can be associated to an observable, a Hermitian operator $O$ acting on the Hilbert space of the system. von Neumann's first assumption is that this correspondence is one-to-one and onto.

**Assumption A.1** *There is a bijective correspondence between measurements in a quantum system and Hermitian operators acting on the Hilbert space of the system.*

This is not always the case, since some systems are subjected to certain superselection rules, which forbid some Hermitian operators [Wikd]. Although this is not a general assumption, there are other formulations of von Neumann's result that circumvent this difficulty (see [CFS70] and the references therein).

Suppose a completion of quantum theory is given. The states of the system, now corresponding to a quantum state plus an extra variable provided by the completion, define expectation-value functions acting on the set of measurements in the system, which is, by Assumption A.1, the set $O(\mathfrak{H})$ of Hermitian operators acting on the Hilbert space $\mathfrak{H}$ of the system. Then, every state in the completion is associated with an expectation-value function

$$\langle\,\rangle : O(\mathfrak{H}) \longrightarrow \mathbb{R}. \tag{A.5}$$

The next step in von Neumann's approach was to impose a few assumptions on the functions $\langle\,\rangle$ that he believed to be valid if these functions came from expectation values of a real physical system.

**Assumption A.2**

1. *If M is by nature non-negative, then $\langle M \rangle \geq 0$;*
2. *If measurement $M_1$ is associated to observable $O_1$ and $M_2$ is associated to observable $O_2$, we can define measurement $M_1 + M_2$ and it is associated to observable $O_1 + O_2$;*
3. *If $M_1, M_2, \ldots$ are arbitrary measurements, then*

$$\langle a_1 M_1 + a_2 M_2 + \ldots \rangle = a_1 \langle M_1 \rangle + a_2 \langle M_2 \rangle + \ldots, \tag{A.6}$$

*that is, all expectation-value functions are linear;*
4. *If measurement M is associated to observable $O$ and $f : \mathbb{R} \longrightarrow \mathbb{R}$ is any real function,[1] the measurement $f(M)$ is associated to observable $f(O)$.*

**Theorem A.1** *Under Assumptions A.1 and A.2, the expectation-value functions in any theory completing quantum theory are given by*

$$\langle M \rangle = Tr(O\rho), \tag{A.7}$$

---

[1]Measurement $f(M)$ is defined using the following rule: measure M and apply $f$ to the outcome obtained. Observable $f(O)$ can be defined easily if we write $O$ in spectral decomposition. Let $O = \sum_i a_i |v_i\rangle \langle v_i|$, where $\{|v_i\rangle\}$ is an orthonormal basis for the corresponding vector space. Then $f(O) = \sum_i f(a_i) |v_i\rangle \langle v_i|$.

*where $O$ is the observable corresponding to measurement $M$ and $\rho$ is a density operator that depends only on the function $\langle \rangle$ (and not on the particular measurement $M$).*

This result implies that, as long as we impose Assumptions A.1 and A.2, we cannot circumvent the quantum rule for expectation values. As we already know, the pure functions of this form are the ones for which the associated density operator is a one-dimensional projector P and these functions only give deterministic outcomes for a small subset of measurements, namely, the ones for which the subspace in which P projects is an eigenspace of the associated observable. This in turn implies that there is no deterministic function, proving the impossibility of completions compatible with quantum theory.

For a long time, it was generally believed that von Neumann's theorem demonstrated that no deterministic theory reproducing the statistical quantum predictions was possible. In 1966, however, J. Bell published a paper with some serious criticism to one of the requirements made for the expectation-value functions [Bel66]. In von Neumann's argument, he requires expectation-value functions to be linear, which is the case for quantum theory, but there is no physical reason to impose this property for more general theories. In fact, as von Neumann pointed out himself in Ref. [vN55], the sum of measurements $a_1 M_1 + a_2 M_2 + \ldots$ is completely meaningless when the measurements involved are not compatible, since there is no way of constructing, in general, the corresponding experimental set-up to implement it. Bell argued that in the case of incompatible measurements, it is not reasonable to require that the expectation values necessarily reflect the observables' algebraic relationships.

It is a special property of quantum theory that the sum of the corresponding observables corresponds to another allowed measurement (as long as Assumption A.1 is valid), and the fact that the expectation value is linear is a consequence of the mathematical rules of quantum theory and is not enforced by any general physical law. In fact, it is not difficult to provide a completion of quantum theory for a qubit, a quantum system with two-dimensional Hilbert space, which does not satisfy linearity of expectation values.

*Example A.1 (An Example of Completion)* In Ref. [Bel66], Bell showed an example of a completion for a qubit agreeing with quantum theory but violating von Neumann's assumption of linearity. Let $O$ be an operator acting on $\mathbb{C}^2$. Since the Pauli matrices

$$\sigma_x = \begin{bmatrix} 0 & 1 \\ 1 & 0 \end{bmatrix}, \ \sigma_y = \begin{bmatrix} 0 & -i \\ i & 0 \end{bmatrix}, \ \sigma_z = \begin{bmatrix} 1 & 0 \\ 0 & -1 \end{bmatrix} \tag{A.8}$$

and the identity $I$ form a basis to the real vector space of $4 \times 4$ Hermitian operators we can always write $O$ in the form

$$O = a_0 I + a_1 \sigma_x + a_2 \sigma_y + a_3 \sigma_z, \tag{A.9}$$

where $a_i \in \mathbb{R}$.

Denote by $v(O)$ the value associated to measurement $O$ in a completion. If we set $|a\rangle = (a_1, a_2, a_3)$, the eigenvalues of $O$, and hence the possible values of $v(O)$, can be written as

$$v(O) = a_0 \pm \|a\|. \tag{A.10}$$

Let $|\phi\rangle \in \mathbb{C}^2$ and $|n\rangle$ be the point on the Bloch sphere [NC00] corresponding to $|\phi\rangle$. Then, we have

$$\langle O \rangle = \langle\phi| O |\phi\rangle = a_0 + \langle a \mid n \rangle. \tag{A.11}$$

Together with the quantum state $|\phi\rangle$, we will use another vector $|m\rangle$ in the Bloch sphere to represent the state of the system. This new vector plays the role of the extra variable in the completion. The complete state of the system is then given by the pair $(|\phi\rangle, |m\rangle)$, which specifies definite outcomes for every projective measurement according to the rule:

$$\begin{cases} v(O) = a_0 + \|a\| \text{ if } (|m\rangle + |n\rangle) \cdot |a\rangle \geq 0, \\ v(O) = a_0 - \|a\| \text{ if } (|m\rangle + |n\rangle) \cdot |a\rangle < 0, \end{cases} \tag{A.12}$$

in which $v(O)$ is the value assigned to $O$ when the system is in the state $(|\phi\rangle, |m\rangle)$.

This completion returns the quantum statistics when we average over the extra variable $|m\rangle$ using the uniform measure on the sphere $S^2$. Indeed,

$$\int_{S^2} v(O) \, d|m\rangle = \langle O \rangle, \quad \forall |\phi\rangle. \tag{A.13}$$

Hence, Eq. (A.12) defines a completion for a quantum system of two-dimensional Hilbert space. Notice that this assignment is not linear in $O$.

## A.2   Gleason's Lemma

In Ref. [Gle57], Gleason proves his famous theorem, a mathematical result which is of particular importance for the field of quantum logic. It proves that the quantum rule for calculating the probability of obtaining specific results of a given measurement follows naturally from the structure of events in a real or complex Hilbert space. Although Gleason's main result is motivated by a problem in foundations of quantum theory, his main purpose was not related to completions of quantum theory. Nevertheless, one of the corollaries of his main result proves the impossibility of certain completions and it is free of some of the drawbacks present in von Neumann's assumptions.

Gleason's main interest was to determine all measures on the set of subspaces of a Hilbert space.

**Definition A.3** A *measure* in the set $\mathcal{S}$ of subspaces of a Hilbert space $\mathfrak{H}$ is a function

$$\mu : \mathcal{S} \longrightarrow [0, 1] \tag{A.14}$$

such that $\mu(\mathfrak{H}) = 1$ and such that if $\{S_1, \ldots, S_n\}$ is a collection of mutually orthogonal subspaces spanning the subspace $S$

$$\mu(S) = \sum_{i=1}^{n} \mu(S_i). \tag{A.15}$$

*Example A.2* To every density operator acting on $\mathfrak{H}$ corresponds a measure $\mu_\rho$ in $\mathcal{S}$ defined by

$$\mu_\rho(S) = \mathrm{Tr}(\rho P_S) \tag{A.16}$$

where $P_S$ is the projector onto $S$.

Gleason's main result states that all measures on $\mathcal{S}$ are of the form (A.16), if the dimension of $\mathfrak{H}$ is at least three.

**Definition A.4** A *frame function of weight* $W$ for a Hilbert space $\mathfrak{H}$ is a real-valued function

$$f : \mathscr{E} \longrightarrow \mathbb{R} \tag{A.17}$$

where $\mathscr{E}$ is the unit sphere in $\mathfrak{H}$, such that if $\{|x_1\rangle, \ldots, |x_n\rangle\}$ is an orthonormal basis for $\mathfrak{H}$, then

$$\sum_i f(|x_i\rangle) = W. \tag{A.18}$$

Given a non-negative frame function with weight $W = 1$, we can define a measure on $\mathcal{S}$. For every one-dimensional subspace $S$ of $\mathfrak{H}$, we define $\mu(P) = f(|x\rangle)$, where $P$ is the projector over $S$ and $|x\rangle$ is a unit vector belonging to $S$.

**Definition A.5** A frame function is said to be *regular* if there exists a Hermitian operator $T$ acting on $\mathfrak{H}$ such that

$$f(x) = \langle x|T|x\rangle \tag{A.19}$$

for all $x \in \mathscr{E}$.

Before stating his main theorem, Gleason proves several intermediate lemmas, among which is the following:

**Lemma A.1** *Every non-negative frame function on either a real or complex Hilbert space of dimension at least three is regular.*

As a consequence of this lemma, we have Gleason's main result:

**Theorem A.2** *Let $\mu$ be a measure on the set $\mathcal{S}$ of subspaces of a Hilbert space $\mathfrak{H}$ of dimension at least three. Then there exists a density matrix $\rho$ such that $\mu = \mu_\rho$.*

The consequences of Gleason's theorem to the foundations of quantum theory appear clearly if one notice that we can interpret the measure defined not on the set of subspaces, but on the set of corresponding orthogonal projectors. Every projector acting on $\mathfrak{H}$ corresponds to an outcome of a measurement in the corresponding quantum system, and hence a measure on $\mathcal{S}$ defines a way of calculating the probabilities of these outcomes. What Theorem A.2 states is that the only way of defining these probabilities consistently is through the quantum rule using density matrices.

This is certainly a really interesting fact, but for us the most important statement in Gleason's paper is Lemma A.1. This result implies that all measures on $\mathcal{S}$ are *continuous*, and this discards the possibility of certain completions of quantum theory.

## A.2.1   Using Gleason's Lemma to Discard Completions of Quantum Theory

Let $\lambda$ define a deterministic state in a completion of quantum theory describing a system whose associated Hilbert space has dimension at least three. Then, every one-dimensional projector $P$ has a well-defined outcome for $\lambda$ and hence we can define a measure

$$\mu_\lambda : \mathcal{E} \longrightarrow \{0, 1\} \tag{A.20}$$

that takes each vector in $\mathcal{E}$ to the value associated to the projector in this direction by $\lambda$. As a consequence of Lemma A.1, this measure is continuous and hence it has to be a constant function.

To see that this is really the case, we can translate the problem of assigning values to the points of the sphere to a problem of colouring the sphere: if the value associated to a one-dimensional projector is 1, we paint the corresponding unit vectors in green; if the associated value is 0, we paint the vectors in red. Suppose now that there are two vectors with different colours. Then, if we choose a path between the corresponding points in the sphere, we have to change abruptly from red to green somewhere in the way from one point to the other. Hence, the association cannot be done continuously if we use both colours.

Since all associations are constant and we know that, given a pure quantum state, there is at least one unidimensional projector with definite outcome 1, we conclude that for all states in the completion and for all one-dimensional projectors the associated definite value is 1. This clearly cannot reproduce the statistics of quantum theory.

At first sight, one may think that the argument above puts an end to the discussion on the possibility of completions of quantum theory. Although very compelling, there is one extra assumption on the kind of completion considered that was not explicitly mentioned. This extra assumption seems so natural that one may not even realise it is there. Hence, the reasoning above is not enough to discard all kinds of completions of quantum theory. It proves only that the *noncontextual* completions are ruled out.

## A.2.2    The "Hidden" Assumption of Noncontextuality

In the argument used above, it was tacitly assumed that the measurement of an observable must yield the same outcome regardless of what other compatible measurements can be made simultaneously. As we already know, this is the hypothesis of *noncontextuality*. Hence, the completions considered are not general enough, and the argument cannot be used to rule out completely the possibility of completing quantum theory.

With these observations, we can conclude as a corollary of Gleason's lemma the following result:

**Theorem A.3 (Bell-Kochen-Specker)** *There is no noncontextual completion of quantum theory.*

Although this result follows from Gleason's lemma, as we proved above, this fact was noticed only after it was proved by other means by Kochen and Specker. The advantage of Kochen and Specker proof is that, contrary to Gleason's lemma, it uses only a finite number of projectors.

## A.3    Kochen and Specker's Proof

In any completion of quantum theory, observables are assigned a definite value whenever we fix a quantum state for the system and also the value of the extra variable. We will denote this value for observable O by $v(O)$. From now on we will use mainly observables whose associated operators are one-dimensional projectors, since this special case is simpler and is already sufficient to get the desired contradiction and prove the Bell-Kochen-Specker theorem. Since the value assigned to each projector P by the completion must be consistent with quantum theory, $v(P)$ must be one of its eigenvalues, and hence we have

$$v(P) \in \{0, 1\}. \tag{A.21}$$

We also require that the assignment $v$ preserves the algebraic relations among *compatible* operators, and hence, if $P_1, \ldots, P_n$ are orthogonal projectors such that $\sum_i P_i = I$ we demand that

$$\sum_i v(P_i) = 1. \tag{A.22}$$

This means that whenever we have a set of projectors $\{P_i = |\phi_i\rangle \langle\phi_i|\}$ where $\{|\phi_i\rangle\}$ is a basis for the associated Hilbert space $\mathfrak{H}$, there is a single $i'$ for which $v(P_{i'}) = 1$, while $v(P_i) = 0$ for all $i \neq i'$.

Since $v$ comes from a completion of quantum theory, its domain is the set of observables in a quantum system. Nevertheless, we will use the fact that we are restricted to the set of one-dimensional projectors and consider $v$ as function assigning values to either the one-dimensional projectors acting on $\mathfrak{H}$ or the unit vectors in $\mathfrak{H}$. If $P = |\phi\rangle \langle\phi|$, the value of $v$ for both P and $|\phi\rangle$ is the same and we may denote this value by $v(P)$ or $v(\phi)$.

The idea behind Kochen and Specker's proof is to find a set of vectors in such a way that is impossible to assign definite values to the corresponding projectors obeying (A.21) and (A.22). This proves the impossibility of noncontextual completions of quantum theory.

**Definition A.6** A *definite prediction set* of vectors (DPS) is a set $\{|u_1\rangle, \ldots, |u_n\rangle\}$ of unit vectors in a Hilbert space $\mathfrak{H}$ such that for at least one choice of assignment $v(u_i)$ for some $|u_i\rangle$ the value of some other $|u_j\rangle$ is uniquely determined by conditions (A.21) and (A.22).

Such a set may be represented with a graph, usually called *Kochen-Specker diagram*. The vertices of the graph correspond to the vectors in the set and two vertices are connected by an edge if the corresponding vectors are orthogonal. In this representation, the problem of assigning values to the associated projectors can be translated into a problem of colouring the vertices of the graph: if a completion assigns value 1 to the projector, we paint the corresponding vertex in green; if the completion assigns value 0, we paint the vertex in red. Notice that the painting is independent of other compatible measurements performed simultaneously, which is a consequence of the assumption of noncontextuality.

Equation (A.22) implies a rule for the colouring: in a set of mutually orthogonal vectors, at most one can be green; if a set of vectors is an orthogonal basis for $\mathfrak{H}$, one, and only one of them is green.

The DPS used in Kochen and Specker's proof is composed of three-dimensional vectors, with associated diagram shown in Fig. A.1. Such a set is called a KS-8 set.

**Theorem A.4** *The set KS-8 is a DPS.*

*Proof* If vector $|A\rangle$ is green, $|B\rangle$ and $|C\rangle$ must necessarily be red. If $|H\rangle$ is green, $|F\rangle$ and $|G\rangle$ are necessarily red. Since the vectors in the set KS-8 belong to a three-dimensional Hilbert space, vectors $|D\rangle$ and $|E\rangle$ are necessarily green, which is a contradiction since $|D\rangle$ and $|E\rangle$ cannot be green at the same time. Hence, we

**Fig. A.1** The set KS-8. If
$v(A) = 1$, then $v(H) = 0$

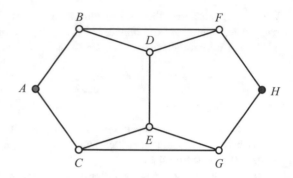

conclude that

$$v(A) = 1 \implies v(H) = 0. \tag{A.23}$$

$\square$

A KS-8 can be constructed using the following vectors in three-dimensional
space:

$$
\begin{aligned}
|A\rangle &= \left(1, 0, 0\right) & |E\rangle &= \left(\tan(\phi)\csc(\beta), -\sin(\beta), \cos(\beta),\right) \\
|B\rangle &= \left(0, \cos(\alpha), \sin(\alpha)\right) & |F\rangle &= \left(\cot(\phi), 1, -\cot(\alpha)\right) \\
|C\rangle &= \left(0, \cos(\beta), \sin(\beta)\right) & |G\rangle &= \left(\cot(\phi), 1, -\cot(\beta)\right) \\
|D\rangle &= \left(\tan(\phi)\csc(\alpha), -\sin(\alpha), \cos(\alpha),\right) & |H\rangle &= \left(\sin(\phi), -\cos(\phi), 0\right).
\end{aligned}
\tag{A.24}
$$

In order to $|D\rangle$ and $|E\rangle$ to be orthogonal, we must have

$$\sin(\alpha)\sin(\beta)\cos(\alpha - \beta) = -\tan^2(\phi). \tag{A.25}$$

It is possible to find $\alpha$ and $\beta$ satisfying this condition whenever

$$|\phi| \leq \arctan\left(\frac{1}{\sqrt{8}}\right). \tag{A.26}$$

Adding two more vectors we get another DPS, called KS-10, whose diagram is
shown in Fig. A.2. In a KS-10, if $|A\rangle$ is green, $|J\rangle$ must necessarily be green. In fact,
$v(A) = 1 \implies v(I) = 0$ and $v(H) = 0$. Since every time we have three mutually
orthogonal vectors one of them must be assigned the value 1, we have $v(J) = 1$.
This set is obtained if we use the vectors in KS-8 of Eq. (A.24) with $|I\rangle = \left(0, 0, 1\right)$
and $|J\rangle = \left(\cos(\phi), \sin(\phi), 0\right)$.

**Definition A.7** A set of vectors $\{|u_1\rangle, \ldots, |u_n\rangle\}$ is called a *partially no-colourable
set* (PNS) if there is at least one choice of assignment $v(u_i)$ to some $|u_i\rangle$ that makes
the assignment of values to the other vectors according to rules (A.21) and (A.22)
impossible.

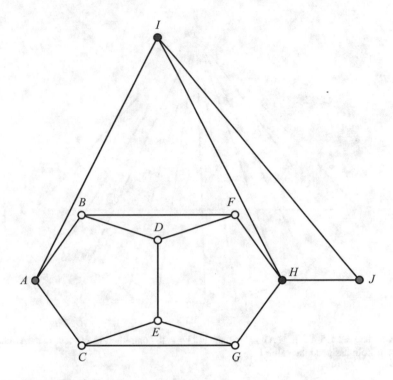

**Fig. A.2** The set KS-10. If $v(A) = 1$, then $v(J) = 1$

To get a PNS we concatenate five diagrams like KS-10, which results in a set of vectors with Kochen-Specker diagram as in Fig. A.3, called KS-42 set.

**Theorem A.5** *The set KS-42 is a PNS.*

*Proof* For this set, the assignment of value 1 to $|A\rangle$ is impossible. In fact,

$$v(\Lambda) = 1 \Rightarrow v(A_1) = 1 \Rightarrow v(A_2) = 1 \Rightarrow v(A_3) = 1 \Rightarrow v(A_4) \qquad (A.27)$$
$$= 1 \Rightarrow v(J) = 1,$$

but $|A\rangle$ and $|J\rangle$ are orthogonal and hence cannot be both green.                        □

**Definition A.8** A set of vectors is called a *totally noncolourable set* (TNS) if it is impossible to assign definite values to all vectors according to rules (A.21) and (A.22).

A TNS provides a proof of the Kochen-Specker Theorem A.3. In fact, a completion compatible with quantum theory must assign values to all projectors (or equivalently, to the corresponding unit vectors) in such a way that conditions (A.21) and (A.22) are obeyed. Hence, if we find a TNS we prove that noncontextual completions consistent with quantum theory are impossible.

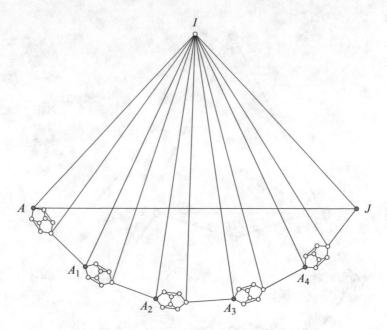

**Fig. A.3** The set KS-42. If $v(A) = 1$, then $v(J) = 1$, contradicting the fact that $|A\rangle$ and $|J\rangle$ cannot be both assigned the value 1

The sphere in any Hilbert space with dimension at least three is a TNS, as we have proven as a corollary of Gleason's lemma. Using three KS-42 sets we can build a TNS with a finite number of vectors in dimension three, simplifying the proof of Theorem A.3. This set is shown in Fig. A.4.

A set of vectors with Kochen-Specker diagram as in Fig. A.4 is called KS-117. This is the set used by Kochen and Specker in their proof of Theorem A.3.

**Theorem A.6** *It is impossible to assign definite values to the vectors of a KS-117 set according to Eqs.* (A.21) *and* (A.22).

*Proof* The vectors $|I\rangle$, $|J\rangle$ and $|K\rangle$ cannot be assigned the value 1, since they are the first vector of a KS-42 set. But they are mutually orthogonal, and hence one of them should be assigned the value 1 according to Eq. (A.22).                     □

The hard part of the proof is to show that there is a set of vectors in a Hilbert space of dimension three with this Kochen-Specker diagram. Actually, this is the reason for concatenating five KS-10 diagrams in order to produce the KS-42: to ensure the representation of such graph with three-dimensional vectors. The list of vectors of the KS-117 and other details can be found in Refs. [KS67, Cab96].

**Fig. A.4** The set KS-117, a
TNS used in the original
proof of the Kochen-Specker
theorem

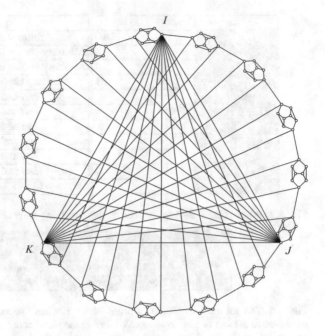

## A.4   Other Additive Proofs of the Bell-Kochen-Specker Theorem

### A.4.1   P-33

One of the simplest proofs of the Bell-Kochen-Specker theorem is due to Asher
Peres and uses a TNS with 33 vectors in a Hilbert space of dimension three [Per91].
This TNS is known as P-33.

To simplify the notation, let $m = -1$ and $s = \sqrt{2}$. The vectors in P-33 are

$$(1, 0, 0), \quad (0, 1, 1), \quad (0, 1, s), \quad (s, 1, 1),$$

$$(0, m, 1), \quad (0, m, s), \quad (s, m, 1), \quad (s, m, m), \tag{A.28}$$

and all others obtained from these by relevant permutations of the coordinates.
By relevant we mean any permutation that generates a vector in a different one-
dimensional subspace, since what is important for the proof is the projector on the
one dimensional subspace and not the vector itself.

The set above has an important property: it is invariant under permutations of the
axis and by a change of orientation in each axis. This allows us to assign value 1
to some vectors arbitrarily, since a different choice is equivalent to this one by an
operation that leaves P-33 invariant.

Figure A.5 shows the proof that P-33 is a TNS. To simplify the notation even
further, we drop parenthesis and commas, using just *abc* to represent the vector

| Trio | | | Vectors ⊥ to the $1st$ | | Explanation |
|---|---|---|---|---|---|
| 001 | 100 | 010 | 110 | $1m0$ | Arbitrary choice of $z$ axis |
| 101 | $m$01 | 010 | | | Arbitrary choice of orientation in $x$ axis |
| 011 | $0m$1 | 100 | | | Arbitrary choice of orientation in $y$ axis |
| 1ms | $m$1$s$ | 110 | $s0m$ | $0s1$ | Arbitrary choice between $x$ and $y$ |
| 10s | $s0m$ | 010 | $smm$ | | $2nd$ and $3rd$ are red |
| s11 | $01m$ | $smm$ | $m0s$ | | $2nd$ and $3rd$ are red |
| s01 | 010 | $10s$ | $mms$ | | $2nd$ and $3rd$ are red |
| 11s | $1m0$ | $11s$ | $0sm$ | | $2nd$ and $3rd$ are red |
| 01s | 100 | $0sm$ | $1s1$ | | $2nd$ and $3rd$ are red |
| 1s1 | $10m$ | $0sm$ | $msm$ | | $2nd$ and $3rd$ are red |
| 100 | $0s1$ | $01s$ | | | CONTRADICTION. |

**Fig. A.5** Proof of the Bell-Kochen-Specker theorem using the set P-33

| 0001 | 0001 | $1m1m$ | $1m1m$ | 0010 | $1mm1$ | $11m1$ | $11m1$ | $111m$ |
|---|---|---|---|---|---|---|---|---|
| 0010 | 0100 | $1mm1$ | 1111 | 0100 | 1111 | $111m$ | $m111$ | $m111$ |
| 1100 | 1010 | 1100 | $10m0$ | 1001 | $100m$ | $1m00$ | 1010 | 1001 |
| $1m00$ | $10m0$ | 0011 | $010m$ | $100m$ | $01m0$ | 0011 | $010m$ | $01m0$ |

**Fig. A.6** The set used in Cabello's proof of the Kochen-Specker theorem using 18 vectors, organised in a table of nine columns. Vectors in the same column are orthogonal. Each vector appears exactly twice

$(a, b, c)$. In the figure, the vectors in each line are mutually orthogonal. The vectors in the first column are assigned the value 1, and hence the other vectors in the same line are assigned the value 0. The assignment of 1 to the vector in the first column is explained in the last column.

We get a contradiction in the last line: we have to assign value 1 to $(1, 0, 0)$, but it is already assigned value 0 in the first line.

In the figure we used only 25 vectors, but we cannot discard the other 8 because we need them to repeat the argument with different choices of the first vector in the first four lines. If we use only the 25 vectors that appear in the figure we would not have a set invariant under permutations of the axis and by change of orientation in each axis, and the set of vectors would not be a TNS.

## A.4.2 Cabello's Proof with 18 Vectors

In 1996, another simple proof of the Bell-Kochen-Specker theorem with 18 vectors in a four-dimensional Hilbert space was found by Cabello et al. [CEGA96]. The TNS in this proof is shown in Figs. A.6 and A.7. Once more, we drop parenthesis and commas to simplify the notation and use $m = -1$. This set provides a simple and elegant proof of the Bell-Kochen-Specker theorem.

**Theorem A.7** *The set of vector shown in Figs. A.6 and A.7 is a TNS.*

**Fig. A.7** Compatibility
hypergraph of the set used in
Cabello's 18-vectors proof of
the Kochen-Specker theorem.
The KS diagram of this set is
the 2-skeleton of this
hypergraph

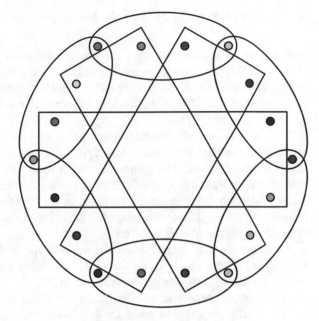

**Fig. A.8** A simplified
version of the KS-diagram.
The dashed edges correspond
to a clique of size 4

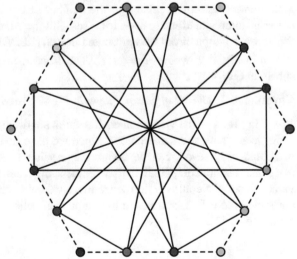

*Proof* In Fig. A.6, the vectors in each column form an orthogonal basis for the four-dimensional Hilbert space. Cells that contain the same vector have the same colour. Since we have nine columns, nine different cells, and only nine, can be assigned the value 1, one for each column. If the assignment is noncontextual, cells with the same colour must be assigned the same value. To see the contradiction, we just notice that the number of cells with the same colour is 2, and hence the number of cells assigned the value 1 must be even (Fig. A.8).                    □

### A.4.3 Cabello's Proof with 21 Vectors

In the previous sections we have seen several different sets that provide a proof of the Bell-Kochen-Specker theorem, and hence the impossibility of noncontextual completions of quantum theory is established. Nevertheless, from a physical point of view, there is still a lot of work to be done. A theorem like this must bring experimental consequences.

The need of an experimental verification of this result is what makes the improvement made by Kochen and Specker's original proof so important: in Gleason's proof, we need an infinite number of vectors to reach a contradiction, and this, of course, makes any experimental test of the result impossible. In the original proof of Kochen and Specker the set of vectors used is finite, but it is still big. Any experimental arrangement involving at least 117 measurements would probably be difficult to implement with small error.

Many proofs were derived after Kochen and Specker's work, with the goal of simplifying as much as possible the TNS used. Among the additive proofs (those relying on Eq. (A.22)), the proof presented in Sect. A.4.2 is still the world record for the smallest number of vectors in the set. But a proof with fewer vectors is not necessarily the simplest proof for an experimentalist. The number of different measurement set-ups is related to the number of contexts, and hence it might be better in some situations to seek for a set with the smallest number of contexts. In this sense, a simple proof was presented in Ref. [LBPC14]. The 21 vectors used are shown in Fig. A.9, where $w = e^{\frac{2\pi i}{3}}$. The Kochen-Specker diagram of this set is shown in Fig. A.10.

**Theorem A.8** *The set of vector shown in Figs. A.9 and A.10 is a TNS.*

*Proof* In Fig. A.9, the vectors in each column are orthogonal. Cells that contain the same vector have the same colour. Since we have seven columns, seven different cells, and only seven, can be assigned the value 1, one for each column. If the assignment is noncontextual, cells with the same colour must be assigned the same value. To see the contradiction, once more we notice that the number of cells with the same colour is 2, and hence the number of cells assigned the value 1 must be even (Fig. A.11).    □

| 100000 | 100000 | 010000 | 001000 | 000100 | 000010 | 000001 |
|---|---|---|---|---|---|---|
| 010000 | 001111 | 001111 | $0101ww^2$ | $0110w^2w$ | $01ww^201$ | $01w^2w10$ |
| 001000 | $0101ww^2$ | $1001w^2w$ | $1001w^2w$ | $1010ww^2$ | $10w^2w01$ | $10ww^210$ |
| 000100 | $0110w^2w$ | $1010ww^2$ | 110011 | 110011 | $ww^20101$ | $w^2w0110$ |
| 000010 | $01ww^201$ | $10w^2w01$ | $ww^20101$ | $w^2w1001$ | $w^2w1001$ | $ww^21010$ |
| 000001 | $01w^2w10$ | $10ww^210$ | $w^2w0110$ | $ww^21010$ | 111100 | 111100 |

**Fig. A.9** The set used in the simplest proof of the Kochen-Specker theorem using 21 vectors and 7 contexts

**Fig. A.10** Compatibility hypergraph of set used in the 21-vector proof of the Bell-Kochen-Specker theorem. Vector labeled by $ij$ is the vector common to $i$-th and $j$-th basis

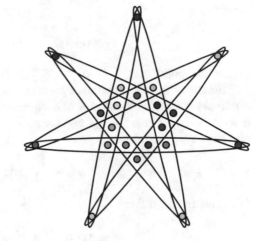

**Fig. A.11** A simplified version of the KS-diagram. Each straight line corresponds to a clique of size 6

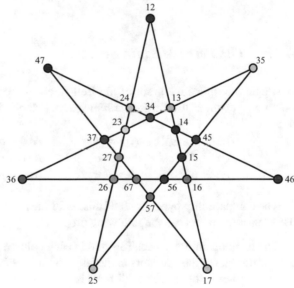

## A.5 Multiplicative Proofs of the Bell-Kochen-Specker Theorem

In the previous proofs of the Bell-Kochen-Specker theorem, we have used the sum of compatible operators and the fact that the values assigned by a completion to the observables should obey the same linear relations the corresponding operators did. More generally, we can assume that, for compatible operators, the validity of

$$f(A_1, \ldots, A_n) = 0 \tag{A.29}$$

implies that

$$f(v(A_1), \ldots, v(A_n)) = 0, \tag{A.30}$$

for any function $f$.

This allows the construction of proofs of the Bell-Kochen-Specker theorem with different functions $f$. Examples of such proofs are the multiplicative ones we will discuss below. In this kind of argument, we use the fact that a set of compatible operators obey the relation

$$A_1 \times \ldots \times A_n = B \tag{A.31}$$

to impose the condition

$$v(A_1) \times \ldots \times v(A_n) = v(B). \tag{A.32}$$

### A.5.1   The Peres Mermin Square

A simple multiplicative proof of the Bell-Kochen-Specker theorem uses the set of operators known as the *Peres-Mermim square* [Mer90, Per90]:

$$\begin{aligned}
A_{11} &= \sigma_x \otimes I & A_{12} &= I \otimes \sigma_x & A_{13} &= \sigma_x \otimes \sigma_x \\
A_{21} &= I \otimes \sigma_y & A_{22} &= \sigma_y \otimes I & A_{23} &= \sigma_y \otimes \sigma_y \\
A_{31} &= \sigma_x \otimes \sigma_y & A_{32} &= \sigma_y \otimes \sigma_x & A_{33} &= \sigma_z \otimes \sigma_z.
\end{aligned} \tag{A.33}$$

The compatibility hypergraph of this set of measurements is shown in Fig. A.12. This set of operators has the following properties:

1. The three operators in each row and in each column are compatible;
2. The product of the operators in the last column is $-I$; the product of the operators in the other columns and in all rows is $I$.

**Theorem A.9** *It is impossible to assign definite values to the measurements in Eq. (A.33) such that $v(A_{ij}) \in \{-1, 1\}$ and such that Eq. (A.32) is satisfied.*

*Proof* Using Eq. (A.32), we have

$$P_1 = v(A_{11})v(A_{12})v(A_{13}) = 1$$

$$P_2 = v(A_{21})v(A_{22})v(A_{23}) = 1$$

$$P_3 = v(A_{31})v(A_{32})v(A_{33}) = 1$$

$$P_4 = v(A_{11})v(A_{21})v(A_{31}) = 1$$

**Fig. A.12** The compatibility hypergraph of the Peres-Mermin square. The product of the measurements in each black hyperedge is the identity operator $I$, while the product of the measurements in the blue hyperedge is equal to $-I$. Vertex in line $i$ and column $j$ corresponds to operator $A_{ij}$

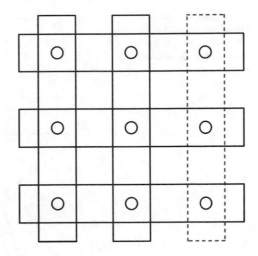

$$P_5 = v(A_{12})v(A_{22})v(A_{32}) = 1$$
$$P_6 = v(A_{13})v(A_{23})v(A_{33}) = -1 \qquad\qquad (A.34)$$

and hence

$$1 = P_1 P_2 P_3 = P_4 P_5 P_6 = -1 \qquad\qquad (A.35)$$

which is a contradiction.                                                      □

This proves that the Peres-Mermim square provides a multiplicative proof of the Bell-Kochen-Specker theorem. The assumption of noncontextuality appears clearly in Eqs. (A.34) since we assumed that each observable has the same value independently of what other compatible observables are measured jointly.

## A.5.2   A Simple Proof in Dimension 8

Another simple multiplicative proof of the Bell-Kochen-Specker theorem was provided by David Mermin in Ref. [Mer90] using the set of operators

$$
\begin{array}{ll}
A_{12} = \sigma_y \otimes I \otimes I & A_{34} = \sigma_x \otimes \sigma_x \otimes \sigma_x \\
A_{13} = \sigma_y \otimes \sigma_y \otimes \sigma_x & A_{23} = \sigma_y \otimes \sigma_x \otimes \sigma_y \\
A_{35} = \sigma_x \otimes \sigma_y \otimes \sigma_y & A_{14} = I \otimes I \otimes \sigma_x \\
A_{25} = I \otimes I \otimes \sigma_y & A_{45} = \sigma_x \otimes I \otimes I \\
A_{15} = I \otimes \sigma_y \otimes I & A_{24} = I \otimes \sigma_x \otimes I.
\end{array}
\qquad (A.36)
$$

**Fig. A.13** Observables
providing a proof of the
Bell-Kochen-Specker
theorem in dimension 8

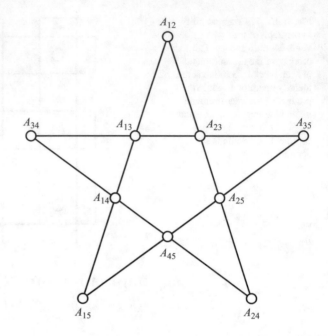

The contradiction we get when we assign definite values to these observables is easily understood if we arrange them in a star, as shown in Fig. A.13. The operators are arranged in five lines with four operators each:

$$A_{12}A_{13}A_{14}A_{15}, \quad A_{12}A_{23}A_{24}A_{45}, \quad A_{13}A_{23}A_{34}A_{35}, \quad A_{14}A_{24}A_{34}A_{45}, \qquad \text{(A.37)}$$
$$A_{15}A_{25}A_{35}A_{45},$$

where $A_{ij}$ denotes the observable belonging to contexts $i$ and $j$. The following properties hold:

1. The observables in each line are compatible;
2. The product of the observables that appear in the horizontal line $A_{13}A_{23}A_{34}A_{35}$ is $-I$; the product of the observables in every other line is $I$.

**Theorem A.10** *It is impossible to assign definite values to the measurements in Eq. (A.36) such that $v\left(A_{ij}\right) \in \{-1, 1\}$ and such that Eq. (A.32) is satisfied.*

*Proof* Using Eq. (A.32), we have

$$P_1 = v(A_{12})v(A_{13})v(A_{14})v(A_{15}) = 1, \qquad \text{(A.38a)}$$

$$P_2 = v(A_{12})v(A_{23})v(A_{24})v(A_{25}) = 1, \qquad \text{(A.38b)}$$

$$P_3 = v(A_{13})v(A_{23})v(A_{34})v(A_{35}) = -1, \qquad \text{(A.38c)}$$

$$P_4 = v(A_{14})v(A_{24})v(A_{34})v(A_{45}) = 1, \qquad \text{(A.38d)}$$

$$P_5 = v(A_{15})v(A_{25})v(A_{35})v(A_{45}) = 1. \qquad \text{(A.38e)}$$

This leads to a contradiction, since the validity of the equations above would imply

$$-1 = P_1 P_2 P_3 P_4 P_5 = \prod_{ij} v(A_{ij})^2 = 1. \tag{A.39}$$

□

## A.6   Yu and Oh's Proof with 13 Vectors

There exist other state-independent proofs with a smaller number of observables. In Ref. [YO12] the authors present a proof of the Kochen-Specker theorem with 13 vectors.

The idea of the proof is quite different from the additive and multiplicative proofs we have shown above. Usually the Bell-Kochen-Specker theorem is proved by finding a set of vectors such that no value assignment satisfying Eq. (A.30) exists. In Ref. [YO12], the authors show a set of 13 vectors for which all value assignments are consistent with this restriction, while being inconsistent with other predictions of quantum theory.

Consider the vectors

$$
\begin{aligned}
|1\rangle &= (1, 0, 0) & |A\rangle &= (-1, 1, 1) \\
|2\rangle &= (0, 1, 0) & |B\rangle &= (1, -1, 1) \\
|3\rangle &= (0, 0, 1) & |C\rangle &= (1, 1, -1) \\
|4\rangle &= (0, 1, -1) & |D\rangle &= (1, 1, 1) \\
|5\rangle &= (1, 0, -1) \\
|6\rangle &= (1, -1, 0) \\
|7\rangle &= (0, 1, 1) \\
|8\rangle &= (1, 0, 1) \\
|9\rangle &= (1, 1, 0).
\end{aligned}
\tag{A.40}
$$

The corresponding orthogonalities are shown in Fig. A.14.

In a noncontextual value assignment, only one of the vectors $|A\rangle$, $|B\rangle$, $|C\rangle$ and $|D\rangle$, can be assigned the value 1. Indeed, suppose that two of these vectors are assigned the value 1. Due to the symmetry of the graph [YO12], we have to consider two cases:

1. $|A\rangle$ and $|B\rangle$ are assigned the value 1. In this case, vectors $|4\rangle$, $|5\rangle$, $|7\rangle$ and $|8\rangle$ are assigned value 0, which in turn implies that $|1\rangle$ and $|2\rangle$ are assigned the value 1, which is not possible.
2. $|A\rangle$ and $|D\rangle$ are assigned the value 1. In this case, vectors $|5\rangle$, $|6\rangle$, $|8\rangle$ and $|9\rangle$ are assigned value 0, which in turn implies that $|2\rangle$ and $|3\rangle$ are assigned the value 1, which is not possible.

Hence, all noncontextual assignments must satisfy

$$v(A) + v(B) + v(C) + v(D) \leq 1, \tag{A.41}$$

**Fig. A.14** Orthogonalities of
Yu and Oh's 13-vectors proof
of the Kochen-Specker
theorem. A noncontextual
completion assigns value 1 to
only one of the blue vertices

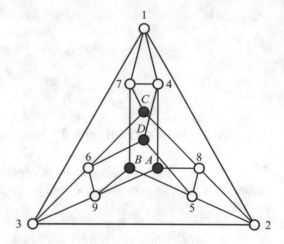

which implies that for all noncontextual completions

$$\langle A \rangle + \langle B \rangle + \langle C \rangle + \langle D \rangle \leq 1. \tag{A.42}$$

However, we have that

$$|A\rangle \langle A| + |B\rangle \langle B| + |C\rangle \langle C| + |D\rangle \langle D| = \frac{4I}{3} \tag{A.43}$$

which implies that for all quantum states

$$\langle A \rangle + \langle B \rangle + \langle C \rangle + \langle D \rangle = \frac{4}{3} > 1. \tag{A.44}$$

This provides an experimentally testable inequality involving only 13 observables
that is satisfied by all noncontextual behaviours while being violated by all qutrit
states. This inequality is, however, not tight. Tight inequalities were derived in
Ref. [KBL+12].

It was shown in Ref. [CKP16] that there is no state-independent contextuality
with less than 13 vectors. This implies that the set of vectors in Eq. (A.40) is actually
a minimal set in quantum theory.

## A.7   A Contextual Completion

The Bell-Kochen-Specker theorem forbids noncontextual completions of quantum
theory, but it is possible to complete quantum theory in order to give definite values
for all projective measurements if we remove the assumption of noncontextuality.
An example of such a completion was provided by Bell in Ref. [Bel66].

To define the completion it suffices to define the values $v(P)$ attributed to the projectors $P$. This happens because every Hermitian operator can be written as a linear combination of compatible projectors

$$A = \sum_i \lambda_i P_i, \tag{A.45}$$

in which $\lambda_i$ is the eigenvalue of $A$ corresponding to eigenspace $P_i$. As we can choose the vectors $P_i$ mutually orthogonal, we can assume that $[P_i, P_j] = 0$ and hence they are mutually compatible. Since the assignment $v$ must preserve the linear relationships between compatible vectors, we have

$$v(A) = \sum_i \lambda_i v(P_i). \tag{A.46}$$

Suppose an experimental arrangement performs the measurement of the observables represented by the projectors $P_1, \ldots, P_n$. Let us define the numbers $a_i \in \mathbb{R}$ such that the expectation values of the $P_1, \ldots, P_n$ are $a_1, a_2 - a_1, a_3 - a_2, \ldots, a_n - a_{n-1}$, respectively. As extra variable in the completion we will use a real number between zero and one. The value associated to each $P_i$ if the value of the extra variable is $\lambda$ is

$$\begin{cases} v(P_i) = 1 \text{ if } a_{i-1} < \lambda \le a_i, \\ v(P_i) = 0 \quad \text{ otherwise.} \end{cases} \tag{A.47}$$

Notice that the value of each $a_i$ depends on the entire set of projectors being measured. Hence the value of $v(P_i)$ does not depend just on the quantum state of the system and the extra variable $\lambda$, it depends also on which other projectors are being measured jointly with $P_i$. This means that this is a contextual completion of quantum theory.

To show that this completion agrees with the quantum predictions, we notice that

$$\langle P_i \rangle = \int_0^1 v(P_i) d\lambda = a_i - a_{i-1}. \tag{A.48}$$

This completion is quite artificial, but it is important conceptually to show that the hypothesis of noncontextuality in the Bell-Kochen-Specker theorem is essential to discard the possibility of completions. It shows that the completions of quantum theory are possible, and brings hope for those who doubt the fact that nature could be intrinsically probabilistic. But one important remark must be made. Completions of quantum theory were first imagined by those who believed that the world could not behave in such a counter-intuitive manner. The main point was to recover the notion we have in classical theory that every measurement has a definite outcome, that *exists* prior to the measurement and is only *revealed* when the measurement is performed. If we choose to keep this line of thought, the Bell-Kochen-Specker theorem forces contextuality on our theories, which is also a really intriguing feature, absent in classical theories. Hence, if quantum theory is really correct, and so far there is no reason to believe it is not, we have to accept the fact that things are a bit weird and our intuition, modeled by our experience with classical systems, cannot be applied to explain its phenomena.

# References

[AB11]     A. Abramsky, A. Brandenburger, The sheaf-theoretic structure of non-locality and contextuality. New J. Phys. **13**, 113036 (2011)

[AMB11]    S. Abramsky, S. Mansfield, R.S. Barbosa, The cohomology of non-locality and contextuality, in *Proceedings 8th International Workshop on Quantum Physics and Logic, QPL 2011* (2011), pp. 1–14

[ABK$^+$15]   S. Abramsky, R.S. Barbosa, K. Kishida, R. Lal, S. Mansfield, Contextuality, cohomology and paradox, in *24th EACSL Annual Conference on Computer Science Logic, CSL 2015, September 7–10, 2015, Berlin, Germany* (2015), pp. 211–228

[ABM17]    S. Abramsky, R.S. Barbosa, S. Mansfield, Contextual fraction as a measure of contextuality. Phys. Rev. Lett. **119**, 050504 (2017)

[AFLS15]   A. Acín, T. Fritz, A. Leverrier, A.B. Sainz, A combinatorial approach to nonlocality and contextuality. Commun. Math. Phys. **334**(2), 533–628 (2015)

[ABB$^+$17]   A. Acín, I. Bloch, H. Buhrman, T. Calarco, C. Eichler, J. Eisert, D. Esteve, N. Gisin, S.J. Glaser, F. Jelezko, S. Kuhr, M. Lewenstein, M.F. Riedel, P.O. Schmidt, R. Thew, A. Wallraff, I. Walmsley, F.K. Wilhelm, The European quantum technologies roadmap. ArXiv:1712.03773 (2017)

[Ama14]    B. Amaral, The exclusivity principle and the set o quantum distributions. Ph.D. Thesis, Universidade Federal de Minas Gerais, 2014

[AT17]     B. Amaral, M.T. Cunha, On geometrical aspects of the graph approach to contextuality. ArXiv:1709.04812 (2017)

[ATC14]    B. Amaral, M.T. Cunha, A. Cabello, Exclusivity principle forbids sets of correlations larger than the quantum set. Phys. Rev. A **89**, 030101 (2014)

[ACTA17]   B. Amaral, A. Cabello, M.T. Cunha, L. Aolita, Noncontextual wirings. ArXiv:1705.07911 (2017)

[ADO17]    B. Amaral, C. Duarte, R.I. Oliveira, Necessary conditions for extended noncontextuality in general sets of random variables. ArXiv:1710.01318 (2017)

[Ara12]    M. Araújo, Quantum realism and quantum surrealism. Master's thesis, Physics, Universidade Federal de Minas Gerais, 2012

[AQB$^+$13]   M. Araújo, M.T. Quintino, C. Budroni, M.T. Cunha, A. Cabello, All noncontextuality inequalities for the $n$-cycle scenario. Phys. Rev. A **88**, 022118 (2013)

[Asp75]    A. Aspect, Proposed experiment to test separable hidden-variable theories. Phys. Lett. A **54**(2), 117–118 (1975)

© The Author(s), under exclusive licence to Springer Nature Switzerland AG 2018          125
B. Amaral, M. Terra Cunha, *On Graph Approaches to Contextuality and their Role in Quantum Theory*, SpringerBriefs in Mathematics, https://doi.org/10.1007/978-3-319-93827-1

[AG16]      A. Auffèves, P. Grangie, Recovering the quantum formalism from physically
            realist axioms. Sci. Rep. **7**, 43365 (2016)
[AII06]     D. Avis, H. Imai, T. Ito, On the relationship between convex bodies related to
            correlation experiments with dichotomic observables. J. Phys. A Math. Gen.
            **39**(36), 11283 (2006)
[Bal13]     P. Ball, Physics: quantum quest. Nature **501**, 154–156 (2013)
[BJ87]      L.E. Ballentine, J.P. Jarrett, Bell's theorem: does quantum mechanics contradict
            relativity? Am. J. Phys. **55**(8), 696–701 (1987)
[BM86]      F. Barahona, A.R. Mahjoub, On the cut polytope. Math. Program. **36**(2),
            157–173 (1986)
[Bar14]     R.S. Barbosa, On monogamy of non-locality and macroscopic averages:
            examples and preliminary results, in *Proceedings 11th Workshop on Quantum
            Physics and Logic*, ed. by B. Coecke, I. Hasuo, P. Panangaden. Electronic
            Proceedings in Theoretical Computer Science, vol. 172 (Open Publishing
            Association, Den Haag, 2014), pp. 36–55
[BW16]      H. Barnum, A. Wilce, *Post-Classical Probability Theory* (Springer, Dordrecht,
            2016), pp. 367–420
[Bar07]     J. Barrett, Information processing in generalized probabilistic theories. Phys.
            Rev. A **75**, 032304 (2007)
[Bel66]     J.S. Bell, On the problem of hidden variables in quantum mechanics. Rev. Mod.
            Phys. **38**, 447–452 (1966)
[BBC+93]    C.H. Bennett, G. Brassard, C. Crépeau, R. Jozsa, A. Peres, W.K. Wooters,
            Teleporting an unknown quantum state via dual classical and EPR channels.
            Phys. Rev. Lett. **70**, 1895 (1993)
[Ber61]     C. Berge, Färbung von graphen, deren sämtliche bzw. deren ungerade kreise
            starr sind, vol. 10 (Wiss. Z. Martin-Luther-Univ. Halle-Wittenberg Math.-
            Natur. Reihe, 1961)
[Bol98]     B. Bollobas, *Modern Graph Theory*. Graduate Texts in Mathematics, vol. 184
            (Springer, Berlin, 1998)
[BCdA+14]   G. Borges, M. Carvalho, P.L. de Assis, J. Ferraz, M. Araújo, A. Cabello,
            M.T. Cunha, S. Pádua, Experimental test of the quantum violation of the
            noncontextuality inequalities for the *n*-cycle scenario. Phys. Rev. A **89**, 052106
            (2014)
[BBL+06]    G. Brassard, H. Buhrman, N. Linden, A.A. Méthot, A. Tapp, F. Unger, Limit
            on nonlocality in any world in which communication complexity is not trivial.
            Phys. Rev. Lett. **96**, 250401 (2006)
[BAC17]     S.G.A. Brito, B. Amaral, R. Chaves, Quantifying Bell non-locality with the
            trace distance. ArXiv: 1709.04260 (2017)
[JBVR17]    D.E. Browne, C. Okay, J. Bermejo-Vega, N. Delfosse, R. Raussendorf, Con-
            textuality as a resource for models of quantum computation with qubits.
            Phys. Rev. Lett. **119**, 120505 (2017)
[BM10]      C. Budroni, G. Morchio, The extension problem for partial Boolean structures
            in quantum mechanics. J. Math. Phys. **51**(12), 122205 (2010)
[Cab96]     A. Cabello, Pruebas algebraicas de imposibilidad de variables ocultas en
            Mecánica Cuántica. Ph.D. Thesis, Universidad Complutense de Madrid, 1996
[Cab13a]    A. Cabello, Proposed experiment to exclude higher-than-quantum violations of
            the Bell inequality. Arxiv:1303.6523 (2013)
[Cab13b]    A. Cabello, Simple explanation of the quantum violation of a fundamental
            inequality. Phys. Rev. Lett. **110**, 060402 (2013)
[CEGA96]    A. Cabello, J.M. Estebaranz, G. García-Alcaine, Bell-Kochen-Specker theo-
            rem: a proof with 18 vectors. Phys. Lett. A **212**, 183–187 (1996)
[CSW10]     A. Cabello, S. Severini, A. Winter, (Non-)contextuality of physical theories as
            an axiom. Arxiv:1010.2163 (2010)

[CBTCB13]   A. Cabello, P. Badziąg, M.T. Cunha, M. Bourennane, Simple hardy-like proof of quantum contextuality. Phys. Rev. Lett. **111**, 180404 (2013)

[CDLP13]   A. Cabello, L.E. Danielsen, A.J. López-Tarrida, J.R. Portillo, Basic exclusivity graphs in quantum correlations. Phys. Rev. A **88**, 032104 (2013)

[CSW14]   A. Cabello, S. Severini, A. Winter, Graph-theoretic approach to quantum correlations. Phys. Rev. Lett. **112**, 040401 (2014)

[CKP16]   A. Cabello, M. Kleinmann, J.R. Portillo, Quantum state-independent contextuality requires 13 rays. J. Phys. A Math. Theor. **49**(38), 38LT01 (2016)

[CAC+16]   G. Cañas, E. Acuña, J. Cariñe, J.F. Barra, E.S. Gómez, G.B. Xavier, G. Lima, A. Cabello, Experimental demonstration of the connection between quantum contextuality and graph theory. Phys. Rev. A **94**, 012337 (2016)

[CFS70]   V. Capasso, D. Fortunato, F. Selleri, von Neumann's theorem and hidden variable models. Riv. Nuovo Cimento **II**(2), 149–199 (1970)

[CDP11]   G. Chiribella, G.M. D'Ariano, P. Perinotti, Informational derivation of quantum theory. Phys. Rev. A **84**, 012311 (2011)

[CRST06]   M. Chudnovsky, N. Robertson, P. Seymour, R. Thomas, The strong perfect graph theorem. Ann. Math. **164**, 51–229 (2006)

[CHSH69]   J.F. Clauser, M.A. Horne, A. Shimony, R.A. Holt, Proposed experiment to test local hidden-variable theories. Phys. Rev. Lett. **23**, 880–884 (1969)

[CGL+02]   D. Collins, N. Gisin, N. Linden, S. Massar, S. Popescu, Bell inequalities for arbitrarily high-dimensional systems. Phys. Rev. Lett. **88**, 040404 (2002)

[DvB11]   B. Dakić, Č. Brukner, Quantum theory and beyond: is entanglement special? in *Deep Beauty, Understanding the Quantum World Through Mathematical Innovation* (Cambridge University Press, Cambridge, 2011)

[DAGBR15]   N. Delfosse, P.A. Guerin, J. Bian, R. Raussendorf, Wigner function negativity and contextuality in quantum computation on rebits. Phys. Rev X **5**, 021003 (2015)

[DL97]   M.M. Deza, M. Laurent, *Geometry of Cuts and Metrics*. Algorithms and Combinatorics, vol. 15 (Springer, Berlin, 1997)

[EPR35]   A. Einstein, B. Podolsky, N. Rosen, Can quantum-mechanical description of physical reality be considered complete? Phys. Rev. **47**, 777–780 (1935)

[Fin82]   A. Fine, Hidden variables, joint probability, and the Bell inequalities. Phys. Rev. Lett. **48**, 291–295 (1982)

[FSA+13]   T. Fritz, A.B. Sainz, R. Augusiak, J.B. Brask, R. Chaves, A. Leverrier, A. Acín, Local orthogonality as a multipartite principle for quantum correlations. Nat. Commun. **4**, 2263 (2013)

[FS16]   C.A. Fuchs, B.C. Stacey, *Some Negative Remarks on Operational Approaches to Quantum Theory* (Springer, Dordrecht, 2016), pp. 283–305

[GVW+15]   M. Giustina, M.A.M. Versteegh, S. Wengerowsky, J. Handsteiner, A. Hochrainer, K. Phelan, F. Steinlechner, J. Kofler, J.-Å. Larsson, C. Abellán, W. Amaya, V. Pruneri, M.W. Mitchell, J. Beyer, T. Gerrits, A.E. Lita, L.K. Shalm, S.W. Nam, T. Scheidl, R. Ursin, B. Wittmann, A. Zeilinger, Significant-loophole-free test of bell's theorem with entangled photons. Phys. Rev. Lett. **115**, 250401 (2015)

[Gle57]   A. Gleason, Measures on the closed subspaces of a Hilbert space. J. Math. Mech. **6**(6), 885–893 (1957)

[GS01]   G. Grimmett, D. Stirzaker, *Probability and Random Processes*. Probability and Random Processes (OUP, Oxford, 2001)

[VVE12]   D. Gross, V. Veitch, C. Ferrie, J. Emerson, Negative quasi-probability as a resource for quantum computation. New J. Phys. **14**, 113011 (2012)

[GLS81]   M. Grötschel, L. Lovász, A. Schrijver, The ellipsoid method and its consequences in combinatorial optimization. Combinatorica **1**, 169 (1981)

[GLS86]   M. Grötschel, L. Lovász, A. Schrijver, Relaxations of vertex packing. J. Comb. Theory Ser. B **40**(3), 330–343 (1986)

[GLS93]      M. Grötschel, L. Lovász, A. Schrijver, *Geometric Algorithms and Combinato-rial Optimization*. Algorithms and Combinatorics (Springer, Berlin, 1993)

[GHH⁺14]     A. Grudka, K. Horodecki, M. Horodecki, P. Horodecki, R. Horodecki, P. Joshi, W. Kłobus, A. Wójcik, Quantifying contextuality. Phys. Rev. Lett. **112**, 120401 (2014)

[Har92]      L. Hardy, Quantum mechanics, local realistic theories, and lorentz-invariant realistic theories. Phys. Rev. Lett. **68**, 2981–2984 (1992)

[Har93]      L. Hardy, Nonlocality for two particles without inequalities for almost all entangled states. Phys. Rev. Lett. **71**, 1665–1668 (1993)

[Har01]      L. Hardy, Quantum theory from five reasonable axioms. arxiv:0101012 (2001)

[Har11]      L. Hardy, Reformulating and reconstructing quantum theory. arxiv:1104.2066 (2011)

[HLB⁺06]     Y. Hasegawa, R. Loidl, G. Badurek, M. Baron, H. Rauch, Quantum contextual-ity in a single-neutron optical experiment. Phys. Rev. Lett. **96**, 230401 (2006)

[HBD⁺15]     B. Hensen, H. Bernien, A.E. Dreau, A. Reiserer, N. Kalb, M.S. Blok, R.F.L. Ruitenberg, J. Vermeulen, R.N. Schouten, C. Abellan, W. Amaya, V. Pruneri, M.W. Mitchell, M. Markham, D.J. Twitchen, D. Elkouss, S. Wehner, T.H. Taminiau, R. Hanson, Loophole-free bell inequality violation using electron spins separated by 1.3 kilometres. Nature **526**, 682–686 (2015)

[HKB⁺16]     B. Hensen, N. Kalb, M.S. Blok, A.E. Dréau, A. Reiserer, R.F.L. Vermeulen, R.N. Schouten, M. Markham, D.J. Twitchen, K. Goodenough, D. Elkouss, S. Wehner, T.H. Taminiau, R. Hanson, Loophole-free bell test using electron spins in diamond: second experiment and additional analysis. Sci. Rep. **6**, 30289 (2016)

[HHHH09]     R. Horodecki, P. Horodecki, M. Horodecki, K. Horodecki, Quantum entangle-ment. Rev. Mod. Phys. **81**, 865 (2009)

[HGJ⁺15]     K. Horodecki, A. Grudka, P. Joshi, W. Kłobus, J. Łodyga, Axiomatic approach to contextuality and nonlocality. Phys. Rev. A **92**, 032104 (2015)

[HLZ⁺09]     Y. Huang, D. Li, M. Cao, C. Zhang, Y. Zhang, B. Liu, C. Li, G. Guo, Experimental test of state-independent quantum contextuality of an indivisible quantum system. Phys. Rev. A **87**, 052133 (2009)

[HWVE14]     M. Howard, J. Wallman, V. Veitch, J. Emerson, Contextuality supplies the /'magic/' for quantum computation. Nature **510**, 351– 355 (2014)

[IKM09]      T. Ito, H. Kobayashi, K. Matsumoto, Oracularization and two-prover one-round interactive proofs against nonlocal strategies, in *2009 24th Annual IEEE Conference on Computational Complexity* (2009), pp. 217–228

[Jam81]      B.R. James, *Probabilidade: um curso em nível intermediário* (Projeto Euclides, Redwood City, 1981)

[KZG⁺09]     G. Kirchmair, F. Zähringer, R. Gerritsma, M. Kleinmann, O. Gühne, A. Cabello, R. Blatt, C.F. Roos, State-independent experimental test of quan-tum contextuality. Nature (London) **460**, 494 (2009)

[KBL⁺12]     M. Kleinmann, C. Budroni, J.-Å. Larsson, O. Gühne, A. Cabello. Optimal inequalities for state-independent contextuality. Phys. Rev. Lett. **109**, 250402 (2012)

[KCBS08]     A.A. Klyachko, M.A. Can, S. Binicioğlu, A.S. Shumovsky, Simple test for hidden variables in spin-1 systems. Phys. Rev. Lett. **101**, 020403 (2008)

[Knu94]      D. Knuth, The sandwich theorem. Electron. J. Comb. **1**, A1 (1994). http://www. combinatorics.org/ojs/index.php/eljc/issue/view/Volume1

[Koc15]      S. Kochen, A reconstruction of quantum mechanics. Found. Phys. **45**(5), 557–590 (2015)

[KS67]       S. Kochen, E. Specker, The problem of hidden variables in quantum mechanics. J. Math. Mech. **17**(1), 59–87 (1967)

[KDL15] J.V. Kujala, E.N. Dzhafarov, J.-Å. Larsson, Necessary and sufficient conditions for extended noncontextuality in a broad class of quantum mechanical systems. Phys. Rev. Lett. **115**, 150401 (2015)

[LMZNCAH18] F.M. Leupold, M. Malinowski, C. Zhang, V. Negnevitsky, A. Cabello, J. Alonso, J.P. Home, Sustained state-independent quantum contextual correlations from a single ion. Phys. Rev. Lett. **120**(18), 180401 (2018). https://doi.org/10.1103/PhysRevLett.120.180401

[LSW11] Y.-C. Liang, R.W. Spekkens, H.M. Wiseman, Specker's parable of the overprotective seer: a road to contextuality, nonlocality and complementarity. Phys. Rep. **506**(1), 1–39 (2011)

[LPSW07] N. Linden, S. Popescu, A.J. Short, A. Winter, Quantum nonlocality and beyond: limits from nonlocal computation. Phys. Rev. Lett. **99**, 180502 (2007)

[LBPC14] P. Lisoněk, P. Badziag, J.R. Portillo, A. Cabello, Kochen-Specker set with seven contexts. Phys. Rev. A **89**, 042101 (2014)

[Lov79] L. Lovász, On the Shannon capacity of a graph. IEEE Trans. Inf. Theory **25**(1), 1–7 (1979)

[Lov95] L. Lovász, Semidefinite programs and combinatorial optimization (lecture notes), 1995

[MM92] S. MacLane, I. Moerdijk, *Sheaves in Geometry and Logic: A First Introduction to Topos Theory*. Mathematical Sciences Research Institute Publications (Springer, New York, 1992)

[Mac98] S. MacLane, *Categories for the Working Mathematician*. Graduate Texts in Mathematics (Springer, Berlin, 1998)

[Man13] S. Mansfield, The mathematical structure of non-locality and contextuality. D.Phil. Thesis, Oxford University, 2013

[MM11] L. Masanes, M.P. Müller, A derivation of quantum theory from physical requirements. New J. Phys. **13**, 063001 (2011)

[Mer89] N.D. Mermin, What's wrong with this pillow? Phys. Today **42**, 9 (1989)

[Mer90] N.D. Mermin, Simple unified form for the major no-hidden-variables theorems. Phys. Rev. Lett. **65**, 3373–3376 (1990)

[Mer14] N.D. Mermin, Physics: QBism puts the scientist back into science. Nature **507**, 421–423 (2014)

[LLS11] R. Lapkiewicz, P. Li, C. Schaeff, N.K. Langford, S. Ramelow, M. Wieśniak, A. Zeilinger, Experimental non-classicality of an indivisible quantum system. Nature (London) **474**, 490 (2011)

[NW09] M. Navascués, H. Wunderlich, A glance beyond the quantum model. Proc. R. Soc. **466**, 881 (2009)

[NPA07] M. Navascués, S. Pironio, A. Acín, Bounding the set of quantum correlations. Phys. Rev. Lett. **98**, 010401 (2007)

[NPA08] M. Navascués, S. Pironio, A. Acín, A convergent hierarchy of semidefinite programs characterizing the set of quantum correlations. New J. Phys. **10**(7), 073013 (2008)

[NGHA15] M. Navascués, Y. Guryanova, M.J. Hoban, A. Acín, Almost quantum correlations. Nat. Commun. **6**, 6288 (2015)

[NBD+13] M. Nawareg, F. Bisesto, V. D'Ambrosio, E. Amselem, F. Sciarrino, M. Bourennane, A. Cabello, Bounding quantum theory with the exclusivity principle in a two-city experiment. ArXiv:1311.3495 (2013)

[NC00] M.A. Nielsen, I.L. Chuang, *Quantum Computation and Quantum Information* (Cambridge University Press, Cambridge, 2000)

[PV09] K.F. Pál, T. Vértesi, Concavity of the set of quantum probabilities for any given dimension. Phys. Rev. A **80**, 042114 (2009)

[PV10] K.F. Pál, T. Vértesi, Maximal violation of a bipartite three-setting, two-outcome Bell inequality using infinite-dimensional quantum systems. Phys. Rev. A **82**, 022116 (2010)

[PPK+09]    M. Pawłowski, T. Paterek, D. Kaszlikowski, V. Scarani, A. Winter,
            M. Żukowski, Information causality as a physical principle. Nature **461**, 1101
            (2009)

[Per90]     A. Peres, Incompatible results of quantum measurements. Phys. Lett. A
            **151**(3–4), 107–108 (1990)

[Per91]     A. Peres, Two simple proofs of the Kochen-Specker theorem. J. Phys. A Math.
            Gen. **24**(4), L175–L178 (1991)

[PR94]      S. Popescu, D. Rohrlich, Quantum nonlocality as an axiom. Found. Phys. **24**,
            379 (1994)

[PS93]      W.R. Pulleyblank, B. Shepherd, Formulations for the stable set polytope, in
            *Integer Programming and Combinatorial Optimization*, ed. by G. Rinaldi, L.A.
            Wolsey (Springer, Berlin, 1993), pp. 267–279

[RDLT+14]   R. Rabelo, C. Duarte, A.J. López-Tarrida, M.T. Cunha, A. Cabello, Multigraph
            approach to quantum non-locality. J. Phys. A Math. Theor. **47**(42), 424021
            (2014)

[Rau13]     R. Raussendorf, Contextuality in measurement-based quantum computation.
            Phys. Rev. A **88**, 022322 (2013)

[Roc70]     R.T. Rockafellar, *Convex Analysis*. Princeton Landmarks in Mathematics and
            Physics (Princeton University Press, Princeton, 1970)

[Ros67]     M. Rosenfeld, On a problem of C. E. Shannon in graph theory. Proc. Am. Math.
            Soc. **18**, 315 (1967)

[SBBC13]    M. Sadiq, P. Badziąg, M. Bourennane, A. Cabello, Bell inequalities for the
            simplest exclusivity graph. Phys. Rev. A **87**, 012128 (2013)

[SHP17]     D. Saha, P. Horodecki, M. Pawłowski, State independent contextuality
            advances one-way communication. ArXiv:1708.04751 (2017)

[SS17]      D. Schmid, R.W. Spekkens, Contextual advantage for state discrimination.
            ArXiv:1706.04588 (2017)

[SMSC+15]   L.K. Shalm, E. Meyer-Scott, B.G. Christensen, P. Bierhorst, M.A. Wayne, M.J.
            Stevens, T. Gerrits, S. Glancy, D.R. Hamel, M.S. Allman, K.J. Coakley, S.D.
            Dyer, C. Hodge, A.E. Lita, V.B. Verma, C. Lambrocco, E. Tortorici, A.L.
            Migdall, Y. Zhang, D.R. Kumor, W.H. Farr, F. Marsili, M.D. Shaw, J.A. Stern,
            C. Abellán, W. Amaya, V. Pruneri, T. Jennewein, M.W. Mitchell, P.G. Kwiat,
            J.C. Bienfang, R.P. Mirin, E. Knill, S.W. Nam, Strong loophole-free test of
            local realism. Phys. Rev. Lett. **115**, 250402 (2015)

[SW95]      A. Shiryaev, S.S. Wilson, *Probability*. Graduate Texts in Mathematics
            (Springer, Berlin, 1995)

[Spe]       E.P. Specker, Ernst Specker and the fundamental theorem of quantum mechan-
            ics. Video by A. Cabello. https://vimeo.com/52923835

[Spe60]     E.P. Specker, Die logik nicht gleichzeitig entscheidbarer aussagen. Dialectica
            **14**, 239 (1960)

[Sta]       A. Fine, The Einstein-Podolsky-Rosen argument in quantum theory, in *The
            Stanford Encyclopedia of Philosophy* (Winter 2017 Edition), ed. by E.N. Zalta.
            https://plato.stanford.edu/archives/win2017/entries/qt-epr/

[UZZ+]      M. Um, X. Zhang, J. Zhang, Y. Wang, S. Yangchao, D.L. Deng, L. Duan,
            K. Kim, Experimental certification of random numbers via quantum contex-
            tuality. Sci. Rep. **3**, 1627 (2013)

[vD12]      W. van Dam, Implausible consequences of superstrong nonlocality. Nat.
            Comput. **12**, 9–12 (2012)

[vN55]      J. von Neumman, *Mathematical Foundations of Quantum Mechanics* (Prince-
            ton University Press, Princeton, 1955)

[Vor62]     N.N. Vorob'ev, Consistent families of measures and their extensions. Theory
            Prob. Appl. **7**(2), 147–163 (1962)

[WCaPM03]   S.P. Walborn, M.O. Terra Cunha, S. Pádua, C.H. Monken, Quantum erasure: in quantum mechanics, there are two sides to every story, but only one can be seen at a time. experiments show that "erasing" one allows the other to appear. Am. Sci. **91**(4), 336–343 (2003)

[Whe78]   J.A. Wheeler, The "past" and the "delayed-choice" double-slit experiment, in *Mathematical Foundations of Quantum Theory*, ed. by A.R. Marlow (Academic Press, London, 978), pp. 9–48

[Wika]   Wikipedia, Circulant graph

[Wikb]   Wikipedia, Johnson graph

[Wikc]   Wikipedia, Quantum cryptography

[Wikd]   Wikipedia, Superselection

[Wri78]   R. Wright, The state of the pentagon, a nonclassical example, in *Mathematical Foundations of Quantum Mechanics*, ed. by A.R. Marlow (Elsevier, Amsterdam, 1978)

[Yan13]   B. Yan, Quantum correlations are tightly bound by the exclusivity principle. Phys. Rev. Lett. **110**, 260406 (2013)

[YO12]   A. Yu, C.H. Oh, State-independent proof of Kochen-Specker theorem with 13 rays. Phys. Rev. Lett. **108**, 030402 (2012)

# Index

© The Author(s), under exclusive licence to Springer Nature Switzerland AG 2018          133
B. Amaral, M. Terra Cunha, *On Graph Approaches to Contextuality
and their Role in Quantum Theory*, SpringerBriefs in Mathematics,
https://doi.org/10.1007/978-3-319-93827-1

Printed in the United States
By Bookmasters